# CELLULAR MOBILE RADIOTELEPHONES

# CELLULAR MOBILE RADIOTELEPHONES

Stephen W. Gibson

**PRENTICE-HALL, INC.**

*Englewood Cliffs, New Jersey 07632*

*Library of Congress Cataloging-in-Publication Data*

GIBSON, STEPHEN W.
Cellular mobile radiotelephones.

Includes index.
1. Cellular radio. I. Title.
TK6570.M6G48 1987 621.3845 86-21277
ISBN 0-13-121930-8

Editorial/production supervision and
interior design: Richard Woods
Cover design: Wanda Lubelska
Manufacturing buyer: Gordon Osbourne



Printed in the United States of America

10 9 8 7 6 5 4 3 2

ISBN 0-13-121930-8 025

PRENTICE-HALL INTERNATIONAL (UK) LIMITED, *London*
PRENTICE-HALL OF AUSTRALIA PTY. LIMITED, *Sydney*
PRENTICE-HALL CANADA INC., *Toronto*
PRENTICE-HALL HISPANOAMERICANA, S.A., *Mexico*
PRENTICE-HALL OF INDIA PRIVATE LIMITED, *New Delhi*
PRENTICE-HALL OF JAPAN, INC., *Tokyo*
PRENTICE-HALL OF SOUTHEAST ASIA PTE. LTD., *Singapore*
EDITORA PRENTICE-HALL DO BRASIL, LTDA., *Rio de Janeiro*

# Contents

Preface v

Acknowledgments vii

1 The Cellular Mobile Radiotelephone System 1

*What it is; why it is needed; the technology that makes it possible.*

2 Some Background 11

*A brief history of the telephone; the beginnings of mobile telephone communications; early work with automobiles, trains and airplanes.*

3 The Technology in Detail 25

*How the telephone works; what a cellular system is; how it works; system design; possibilities for expansion; the maintenance of the equipment.*

4 The Heart of the Matter 47

*The guts of the system; the switch; the computer; system configurations.*

5 The Cell Site 61

*What it does; location; the radios; the antennas and supports.*

**6 The Mobile Station** 87

*The handset/control unit; the transceiver/logic unit; the antenna; the portable units; interference.*

**7 The Role of the Regulatory Agencies** 117

*A little history; how the FCC works; the FCC and cellular; the position of the state governments.*

**8 The Economics of the System** 131

*The costs to the user; business/personal uses; who it is for.*

**9 What is Being Done Elsewhere** 137

*Who is big overseas; the English experience; satellites; trains and planes again.*

**10 The Future** 149

*Data handling; encryption and security; cellular in the military.*

**Glossary** 165

**Addendum** 171

**Index** 173

# Preface

The pace of our lives at the present time seems to require that we remain in constant communication, no matter where we are, with those with whom we conduct our business. For a lot of people, this same need applies to their social life—they find it necessary to be in constant touch with the rest of the world.

This ability has been possible for most of us for a long time as long as we were within reach of a telephone at home, in the office, or in a booth on the street. The advent of the submarine cable and now the commercial satellites make it possible for us to contact anyone anywhere in the world with relative ease. Now you can have this capability from your car or boat or while walking your dog. If all goes as planned, you'll be able to have the same capability from a railroad train or commercial aircraft without any difficulty.

There is no doubt that the telephone is one of the essential tools of business life today, and it is a very necessary part of most households. Many houses now being constructed have telephone wiring installed throughout the house while the house is still under construction. This feature is accepted in the same manner as electrical wiring or plumbing. Statistics vary depending on where they come from, but it seems certain that there are more homes in the U.S. now with telephones than there are with inside bathrooms or with washing machines. Also, over half of the homes with telephones have one or more extensions.

Since the early 1920s when the Detroit Police Department was the first government agency to use mobile radiotelephones in their police cars, this capability has been available for some private citizens. Advertising people, when they wanted to illustrate the ultimate aura of luxury, would show a well-dressed businessman in the back of a large limousine with a very glamorous woman, obviously his secretary, handing him a telephone.

Today, ads are somewhat different. The photographic models, location and props show us that this service—a telephone in every car—is now available to every businessperson, real estate broker, or construction superintendent. People can now make telephone calls from their cars with ease.

There is no need to know or worry about the work being done by the maze of electronic circuitry and the tiny components in the mobile telephone that is transmitting the conversation. There is no need to worry about the base station that is receiving it. There is no necessity to be aware that the radio signal radiated from the 9 in. antenna on the trunk or roof of the car is being constantly monitored at the base station. There is no need to know that a computer is measuring the strength of the radio signal, comparing it with the strength of the same signal received through other base stations, and "handing it over" to the station that received the strongest signal.

Furthermore, you need not worry that, while the computer is doing all this, it is also storing away information needed to make out your bill at the end of the month and making a traffic analysis so that the system operator can be sure that proper and profitable service is being maintained by the system.

You need not give this a thought as you continue to talk as you drive. However, if you want to learn a little more about this newest marvel of the electronic age, then read on. You will find out about such things as Standard Metropolitan Statistical Areas (SMSA)—you didn't know that you lived in one, did you?—Cellular Geographic Service Areas (CGSA), Mobile Telephone Switching Offices (MTSO). This is all part of the new jargon that is used to define the cellular mobile radiotelephone communications system that we are going to look at.

In the first chapter of the book we talk very generally about the cellular mobile radiotelephone system, what it is and why it is needed, and we tell you a little bit about the technology that makes it all possible. Chapter 2 gives some history of the telephone along with the background of mobile communications in automobiles, trains, and airplanes.

In Chapter 3 we talk about and describe the technology in some detail and in Chapters 4, 5 and 6, we examine and learn about the various ingredients that make up this method of communication.

The role of the various governmental bodies, federal, state and local, are examined in Chapter 7, and the economics of the system for the user and the system owners are given a broad brush in Chapter 8.

In chapter 9 we tell you what is being done in the rest of the world, and in Chapter 10 we take a look into a crystal ball and talk to some experts to try to see what the future has to offer in this type of communication.

Some of the terms that will be used in this book may be strange to you; they will be kept to a minimum. To make the reading as easy and smooth as possible, a glossary of words, terms, and phrases that may need some explanation will be found at the end of the book.

This is the business that *Fortune* magazine referred to as "The hottest new U.S. industry . . ." in 1984 and went on to predict that there would be a million subscribers with cellular radiotelephones in their cars by 1990. Some of the analysts who study the telecommunications industry are predicting a $12 billion market by 1995. This market, "they" say, will be split three ways, with one-third coming from the sales of the equipment, one-third coming from revenues from the service itself, and the remaining third coming from costs of installing, operating, and servicing the systems.

It is really too soon to be able to predict exactly where the industry is going. The first systems have been operating only since 1983 in this country, and there is really no solid base of data upon which to make judgments.

There is little doubt, however, that the industry is here to stay and that the use of mobile radiotelephones will become increasingly popular in years to come. It is hoped that this book will help you reach an intelligent decision about whether a cellular radiotelephone should be installed in your car.

This book, of course, will not show you how to set up, operate, and maintain a cellular mobile radiotelephone system. Rather, it will explain one more advance in our technological society, help you understand it, and help you decide how a radiotelephone system can help you in your life or your business.

Good luck!!

# ACKNOWLEDGMENTS

In a book that attempts to cover a fairly new industrial and business phenomenon, help must come from a number of people with expertise in differing fields. It is impossible to acknowledge everyone with whom I discussed cellular radiotelephones and from whom I received suggestions, information, literature and photographs. The names and organizations given below deserve special mention, however. They are listed in no particular order. I would like to express my appreciation to everyone who knowingly or otherwise made a contribution in this field. Many thanks!!

Marcia Burgoon and Bland McCartha, Cartelco; Charles Hoovler and his staff at Communications, Inc; Stuart Crump, Benn Kobb and Elaine Lussier at Personal Communications Technology; Rhonda Wickham of Cellular Business, the trade journals that were a big help; Peggy Kingsley, Robert Schied, Pat Schod, Janet Spinks, and Dan Davies of Motorola in Washington, D.C., Schaumburg, IL, and Phoenix, AZ; David Diggs, David Schek, Gary Brunt, and Pamela Bloodworth of Cellular One.

Katherine Boudas and JoAnne Wrenn at Ameritech; Mary Anne Reynolds at the American Automobile Association; Al Lauersdorf of the National Safety Council; Christine Williams at Bell Atlantic; Pat Vance of Broadcasting Magazine; Maxine Carter of OKI Telecommunications; Scott Goldman at Compucom; Lee Hardy with NovAtel; Barbara Hagen at Ericsson; Bob Joiner of CTI; Lori LaJudice at Sinclair; Patrick Mayben of Spectrum Cellular.

Special thanks go to Gareth Davis of Exide; Millie Etlinger, AT&T; and to Robert Maher of the Cellular Telecommunications Industry Association.

For their help with the regulatory side, I want to thank John Gibbons of the National Association of Regulatory Commissioners; and from the FCC, Michael Sullivan of the Personal Radio Bureau and Andrew Nachby who reviewed what I said about the FCC in Chapter 7 and gave me some valuable suggestions.

In the UK I'd like to thank particularly M.C. Wright of the Department of Trade and Industry, D.E. MacKinney and K.J. Hydon of Racal Telecommunications Group, and Peter Carpenter and Ashley Rayfield of Cellnet. In addition, the technical staffs of the French and Japanese Embassies were of help to me.

# CELLULAR MOBILE RADIOTELEPHONES

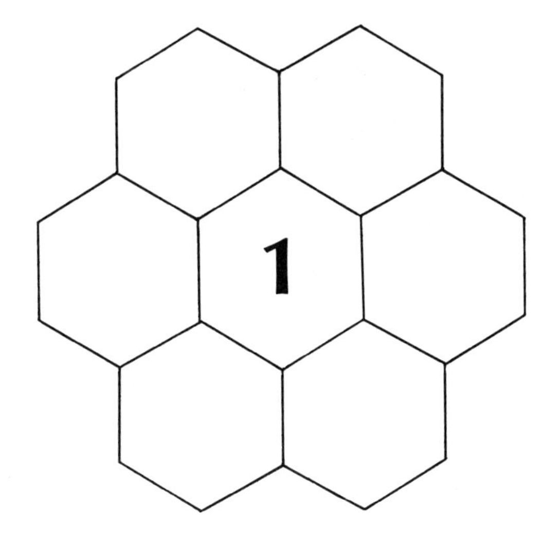


# The Cellular Mobile Radiotelephone System

In very general terms the introduction told a little about this new mobile radiotelephone system and what it could do for businesspeople and the general public. In the chapters that follow, each part of the system is explained in detail.

We begin with a rather long sentence: A cellular mobile radiotelephone system uses a network of relatively low-powered FM transmitters and receivers covering a city or town and the surrounding area and acts as a relay and interface between the portable or mobile radiotelephone in your car or briefcase and the local telephone company. While in use, the system is constantly monitoring the radio transmissions to and from the mobile radiotelephone and, as the vehicle moves, shifting the connection to the ground station that will provide the highest quality communication channel. This is all done automatically with some rather sophisticated electronics and microcomputer programming. The concept is shown schematically in Fig. 1-1. In some systems the Mobile Telephone Switching Office (MTSO) is simply called "the switch" and the local telephone company is usually referred to as the Public Telephone Switched Network (PTSN).

For the past several years there have been ads in the newspapers and magazines, especially those catering to the business community, demonstrating the ease with which telephone calls can be made from a car. The telephone instrument shown in Fig. 1-2 is very much like one

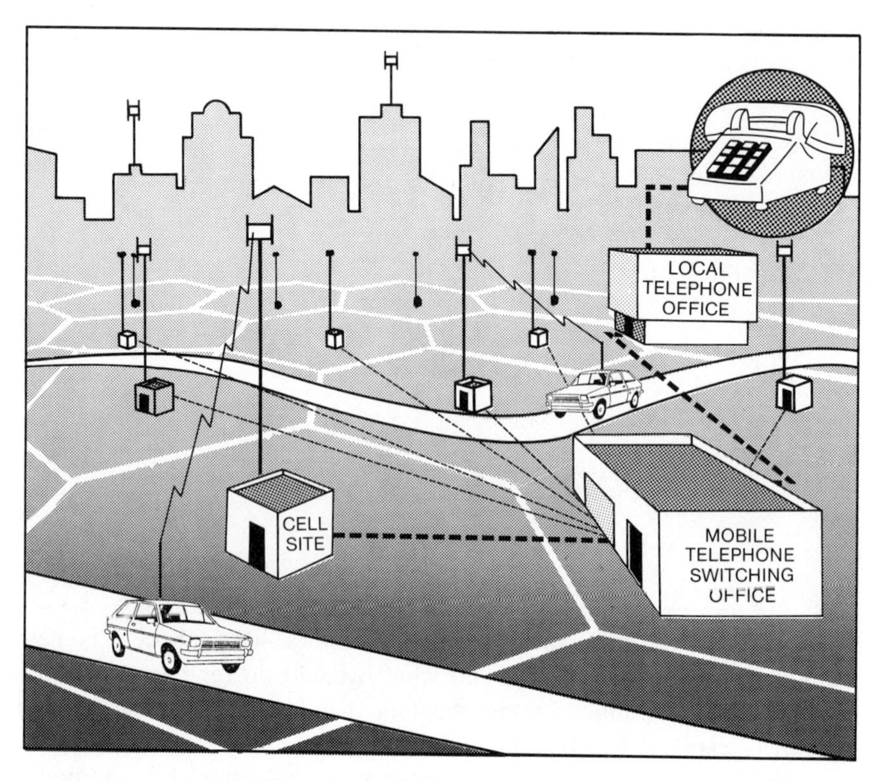


**Figure 1-1.** The cellular concept. (Courtesy of Bell Atlantic Mobile Systems.)

**Figure 1-2.** A cellular mobile telephone in use. (Courtesy of Motorola, Inc.)

installed by the local telephone company or purchased from a local electronics or radio store. The touchtone pad used for *dialing up* the wanted number, however, is on the back of the handset instead of on the front or in the base of the unit. It is true what the ads say—making a telephone call from a vehicle to anywhere in the world is now as simple and easy as making one from a conventional installation.

This was not true in the past with the mobile radiotelephone systems then in use, and the details of the cellular system will show why this new system is superior to the old systems.

Although the emphasis seems to be on the use of the new system in a vehicle, it can be used with a hand-held, battery-powered portable unit as well. Since the drain on the battery is greatest during transmissions, the length of time a unit can be used before the batteries need recharging depends on how much talking is done. With present-day batteries and average use, a portable unit can be used for transmitting for 30 minutes to an hour before the batteries need recharging. This step can be done from a vehicle's electrical system or from a charger plugged into a commercial 110-V supply in the home or office.

Cellular mobile radiotelephone seems to have gotten its name from the fact that the service area is made up of a number of cells, generally shown in a honeycomb fashion, each one containing the needed equipment. This equipment consists of a combined transmitter/receiver (generally referred to as a transceiver) control equipment, and a set of antennas to receive and send the signals to and from the mobile stations. All the equipment at a cell is connected by a telephone line or microwave relay to a central switching and master control unit, the MTSO, that is the interface with the telephone company.

Each of the cells uses a group of radio frequencies or channels sufficiently different from the frequencies in use in the adjoining cells to avoid interference. As a general rule the frequency assignments are not repeated from five to seven cells, but since the number of frequencies is limited, efficient use of the spectrum requires that the frequencies be reassigned to other cells located a sufficient distance away. This frequency reuse is one of the major factors in the success of this system and is a major advance over existing mobile radiotelephone systems.

The coverage from the radio at each of these locations (referred to as "cell sites") is the governing factor in determining the size of the cell. The cells in an area with flat terrain, like some of the cities in the Midwest, will be much larger than the cells in an area with hills and valleys such as Johnstown, Pennsylvania, or San Francisco, for example. In the system in the Washington-Baltimore area, for example, each of the 16 cells in the system at the start-up covers about 210 square miles (Fig. 1-3). One of the advantages of the cellular system is that, if the number of subscribers becomes too large to be handled by the available channels in the existing cell, the cell can be divided into smaller cells. The transmitter power is reduced to cover the smaller area.

The type of radio transmission selected for this system is the same as that used for high-quality commercial radio, FM. This essentially

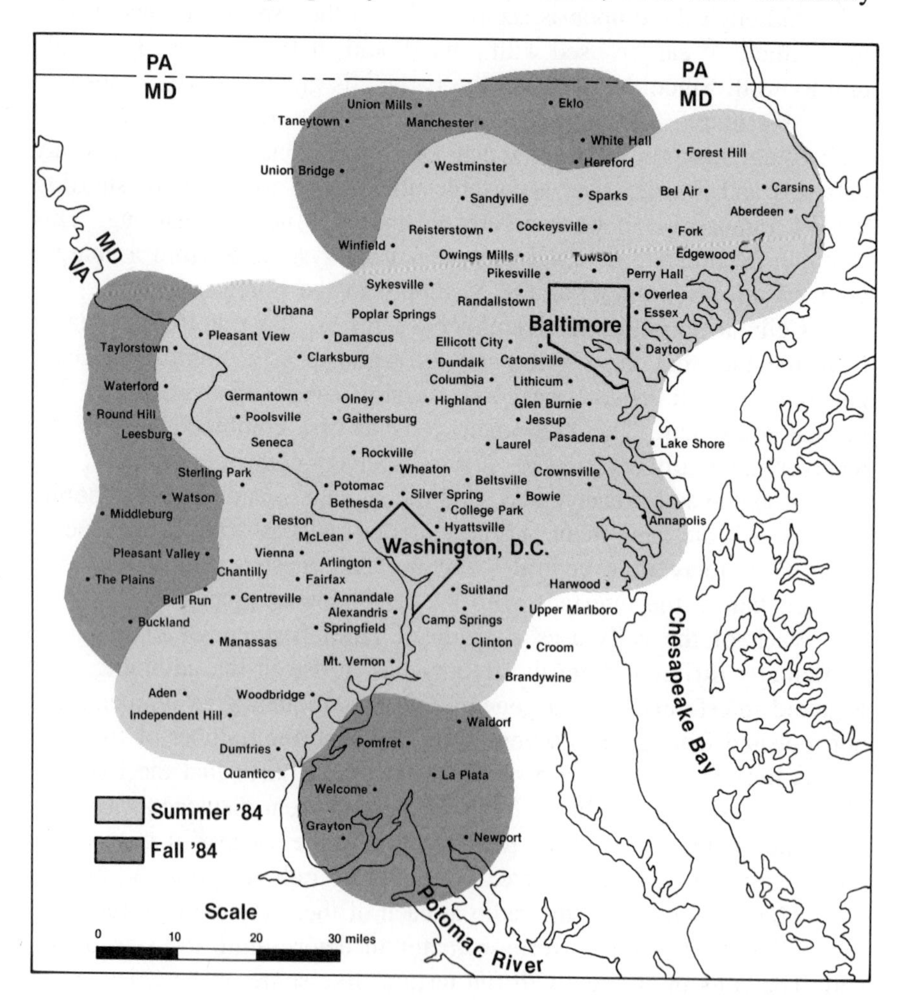


**Figure 1-3.** Coverage of a cellular system at the end of 1985. (Courtesy of Cellular One.)

static and interference-free method of communication helps the "you-could-be-in-the-next-room" effect of the cellular radios.

When people start looking for answers to the question "Why is a new mobile radiotelephone system needed?" they tend to become overwhelmed with numbers and statistics. Let's wade through some of the rough figures that are available to get an idea of the magnitude of the problem that faced the telecommunications industry in the 1960s and 1970s.

Prior to the introduction of the cellular concept, the existing mobile radiotelephone service was designed around the use of a single, high-powered base radio station that covered the area and served the needs of the mobile units in an entire city and surrounding areas. A conventional commercial radio or TV station serves its customers in much the same way, and the number of customers that are served depends, in general, upon how high the antenna can be placed and the amount of power radiated from it.

At that time and with that system, reliable communication with mobile stations was possible within a radius of about 20 miles of the antenna (about 1300 square miles), but the radio signal from the base station even when not strong enough for reliable communication, was strong enough to interfere with the other radio signals on and around that frequency within a radius of about 100 miles.

With this single base station concept, the limitations of the number of frequencies available, and the techniques of the day, there were only about 25 voice channels usable at each station. This meant that in New York, certainly one place in the country where rapid and constant communications are considered a necessity for the financial and corporate community, only about 700 customers could be accommodated. Like the rest of the country, these 700 customers could complete only one out of every two telephone calls the first time they tried. It is easy to see why the market was ready for an improved service.

In 1978, after 20 to 30 years of experimenting, developing, testing, and briefing the regulatory agencies, the first cellular mobile radiotelephone system was commercially tested in Chicago by AT&T and its former subsidiary Illinois Bell. This was approximately 32 years after the introduction of the first commercial mobile radiotelephone service in St. Louis.

Around the country there were about forty-thousand Bell customers with mobile telephones and, as we have said, customers had about a 50% chance of completing a call the first time they tried. In addition, there were about twenty-thousand unhappy potential customers with their names on waiting lists with an expected wait of 5 to 10 years before they would get mobile service because as we've said, of the limitations of the system and equipment then in use.

This brief touch of history shows why the pressure was on the telecommunications industry to get an improved system into operation as quickly as possible. This new mobile radiotelephone system was brought into reality by the same technology that led to the development of calculators the size of a business card, computers of incredible power that can sit on your desk, and solid state electronics—electronic monitoring, control and display of many of the essential operations of the automobile.

The concept of the cellular system had been talked about at the Bell Labs in the 1940s, but the system was not really feasible with the state of technology at the time. Radios of all types were dependent on the vacuum tube with its requirement for large amounts of power and its subsequent emission of large amounts of heat. Computers were filling huge rooms with racks of tubes, relays, power supplies, and other needed accessories. Computer programming of these primitive monsters was in its infancy.

The end of all this came in sight in 1947 with the invention of the transistor, again in the Bell Labs. This led to a great reduction in the size and power requirements for the components and to a tremendous increase in the speed of operation of electronic systems using transistors.

At first engineers and designers treated the transistor purely as a replacement for the vacuum tube in electronic circuits. They continued to use the other components needed to make the circuits —resistors, capacitors, coils—work as individual and separate items. However, as time went on and the engineers learned more and gained experience with the new technology and materials, they found that they could achieve the same results by etching or plating materials on a copperplated sheet of synthetic material to act in the same manner as the previously used individual components. Now the entire circuit could be greatly reduced in size. Circuit boards and chassis with a multitude of components that used to be handled individually and measured in inches on a side could now be measured in tenths of an inch on a side. What once took up a whole chassis and fit on a standard 19 in. electronic rack and panel could now be put on an easily handled circuit board. Figure 1-4 is a greatly enlarged photograph of the Motorola

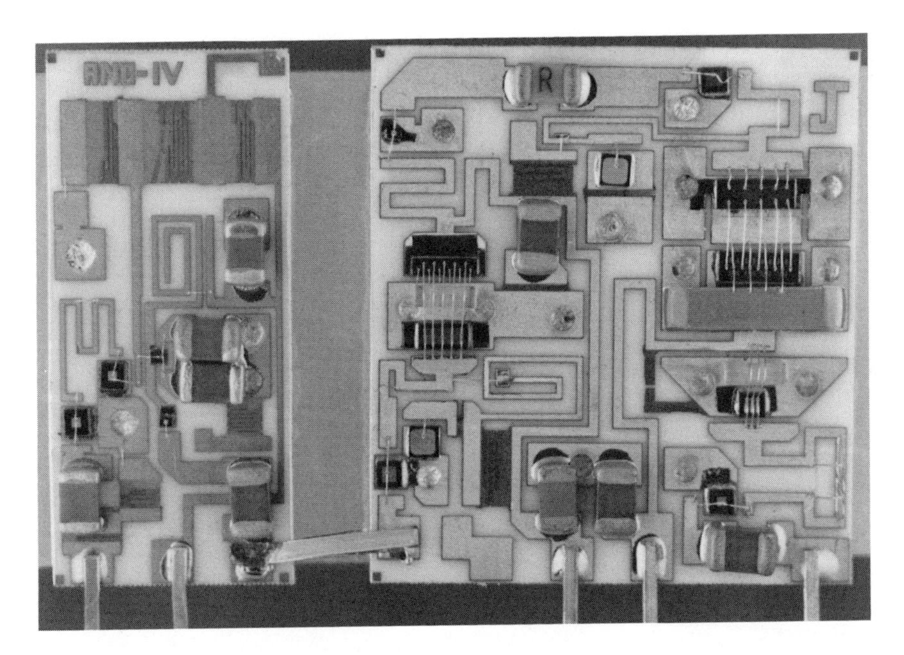

**Figure 1-4.** The MHW Hybrid Amplifier Module manufactured by Motorola. (Courtesy of Motorola, Inc.)

MHW808A1 Hybrid Amplifier Module. This chip contains the components and circuitry to amplify a radio signal to a maximum of $7\frac{1}{2}$ w—about 16 times—and to control the amount of amplification on instructions from a control unit. The chip measures a little over 1 in. long and $\frac{3}{4}$ in. wide. Now the necessary controls became the size-limiting feature—the knobs and switches had to be large enough to be easily handled.

Along with this great reduction in size came a reduction in the length of time that it took the electronic signals to traverse the circuitry. Despite the common impression that electricity travels "at the speed of light," it takes a finite length of time for signals to travel from one end of the circuit to the other. If this time could be reduced with solid state electronics, as it was significantly, the circuit could accommodate more signals.

A particularly important feature for the telephone industry was that signals could now be switched from one circuit to another electronically. There was no need now for any mechanical movements whatsoever, no more clicking relays, to do the multitude of switching functions necessary in an operating radiotelephone system.

The development of the microcomputer—the computer on a chip—and the increased skill of programmers in addition to the benefits of the electronic switch and the Stored Program Control (SPC) all improved the efficiency of the telephone systems and made the cellular mobile radiotelephone system a practical goal.

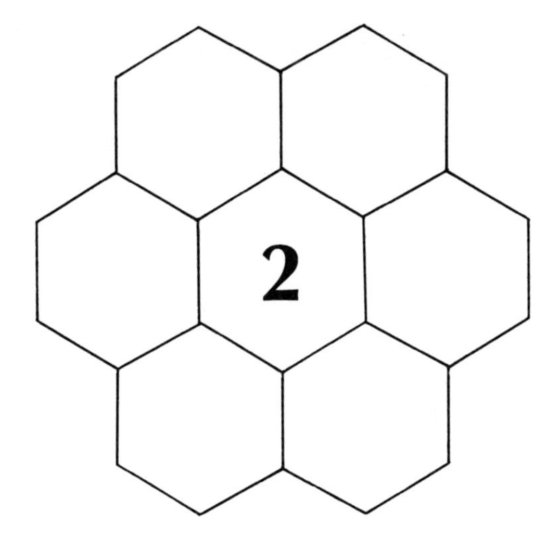


# Some Background

Ever since people devised a spoken language and learned to communicate verbally with others, there has been a search for ways to extend the distance over which messages could carry. A variety of ways has been developed to do this.

Shouting, of course, was probably the first thing tried. Prehistoric people soon learned that they could get their ideas and needs further afield by beating on hollow logs and then on animal skins that were stretched over the ends of logs. Then they learned that they could express more complex ideas by using different rhythms on drums.

For awhile the ancient Greeks used a method of arranging vases in a grid of shelves (Fig. 2-1). Different patterns or numbers of vases meant different things. Incidentally, the word "telephon" has Greek roots and means "far speaking."

Until the discovery of electricity in the late sixteenth century, most methods of extending the range of communication that were tried were visual. Signal fires on hilltops were used for many years—the British spread the news of Wellington's victory at Waterloo up through the island by this method. The smoke signals of the American Indians have been a traditional part of any movie of the Old West.

Semaphores (with movable arms on a pole), as illustrated in Fig. 2-2, were probably the next step. This method had the advantage of making it possible to manipulate the arms by wires—there was no need

**Figure 2-1.** An ancient Greek signalling system.

for the operator to be right at the semaphore site, although the operator had to be able to see the other signals in the chain.

As soon as it was realized that electricity would travel *through* a wire and what could be accomplished by turning the electricity on and off, there was very little further work done to improve visual signalling systems. A lot of work on the use of electricity for communication by a number of inventors and scientists in the U.S. and Europe culminated in May 1844 with Samuel Morse's demonstration of a telegraph system using a wire strung from Washington, D.C. to Baltimore. He sent the words "what God hath wrought" using a coding system that he had devised. This code was made up of dots and dashes created by turning on the power for short and long periods of time.

Before long the nation was crisscrossed by a network of telegraph poles and wires generally paralleling the expanding railroad systems. Many a youth spent hours at the local railroad station watching the

**Figure 2-2.** A signalling semaphore.

telegrapher work magic with the key and listening to the clicks from the telegraph sounder amplified by a Prince Albert tobacco tin jammed behind it.

It was quickly evident, however, that having a third party, the telegraph operator, in the message-handling process was not the ultimate answer. Direct communication by voice on a one-to-one basis soon became the goal. In March 1876 Alexander Graham Bell, during one of a long series of experiments, spilled some acid and, without thinking, used the experimental setup to summon his assistant from the next room with the words "Mr. Watson, come here. I want you."

The value of this new invention was quickly realized, and in 1878 the first telephone switchboard with 38 subscribers and using ex-telegraph boys as operators was installed in New Haven, Connecticut. By 1880 there were 138 exchanges throughout the country. Figure 2-3 shows an early example.

About this same time Guglielmo Marconi, an Italian who was living and working in England, began to put together a number of technical discoveries into a system aimed at eliminating the wires in telegraph systems. By 1895 he had managed to transmit signals through

**Figure 2-3.** An early telephone switchboard in Kansas City, 1904. (Reproduced with permission of AT&T Corporate Archive.)

the air for a distance of two miles on Salisbury Plain, England, without any connecting wires. This was the first *wireless* signal on record.

By 1897 wireless communication had improved to the point where conversations via Morse code were being carried on between shore stations and ships ten miles out at sea. In 1899 Marconi transmitted signals across the English Channel to France. The next great barrier, of course, was the Atlantic Ocean. In 1901, using kites to get his antennas up in the air, Marconi sent the three Morse code dots of the letter *s* from Newfoundland, Canada, to Cornwall, England.

This is the way that it all began, and from then on the story is a mixture of financial manipulation and technical progress. It has been well documented by others from all points of view.

While there were undoubtedly tests and experiments being carried out by the communications industry prior to 1920, the first recorded mobile radiotelephone service was put into use by the Detroit Police Department in that year. Some of the department's scout cars kept in touch with their headquarters with what are now bulky and crude receivers and transmitters.

The first commercial mobile radiotelephone service available to the general public was introduced in St. Louis in 1946. This advance was probably brought about by two happenings during World War II. One event was the great pool of skilled and experienced radio personnel brought about by the increased use of radio communication, both mobile and static, in the armed forces. The other event was the British invention of radar, which was the result of research and development in the higher radio frequencies. Knowledge of this area of the electromagnetic spectrum was invaluable in the postwar development work in mobile radiotelephones.

The system that Bell Telephone installed in St. Louis was assigned channels in the 150-MHz band by the Federal Communication Commission (FCC) but used only three of these bands because of the technical limitations of the time. All the calls to and from the mobile subscribers were routed through a special operator, and the mobile units used a simplex system, what is called "push-to-talk." That is, the speaker talking could not hear the other party on the line. When the speaker had finished that particular part of the conversation and was ready to hear the response, he had to push a switch before hearing the other party.

These early mobile radiotelephone systems all used one high-powered base station per city or service area to blanket the area as described in the previous chapter. This meant that the limited number of channels could not be reused in neighboring cities or towns because the signals would interfere with each other.

Improvements were gradually made in the mobile service as experience was gained in electronics, and the demand grew with the increased use of automobiles for business and pleasure. In 1964 AT&T introduced the Improved Mobile Telephone Service (IMTS), in which the subscriber in the car could dial the wanted number directly on the car instrument—there was no operator intervention as was previously necessary. The available channel selection was done automatically and the waiting time for a clear channel was reduced considerably.

The major problem, lack of capacity, still remained unsolved, and in 1970 the FCC came out with the results of a long study made on the mobile radiotelephone service. This study reviewed in some detail the possibilities of improving the service. The FCC then asked the communications industry to review the results of the study and to submit proposals for a new mobile system using the higher frequencies, the 800-MHz band, that had been made available.

As a result of work that had been going on in the Bell Labs and in response to the FCC request, AT&T filed a proposal for the development of the first cellular system in 1971.

For the next few years, the FCC study and the AT&T proposal that resulted were looked at carefully and in 1974 the FCC allocated 40 MHz of the 800-MHz band for the initial development of commercial radiotelephone systems. In 1975, Illinois Bell filed an application with the FCC proposing to build and test the first cellular system in the Chicago area. This system was to be based upon the Bell Labs work, and in 1977 the FCC gave approval.

The Chicago cellular system, christened Advanced Mobile Phone Service (AMPS), began operation in 1978 using sample subscribers from all segments of the business community to test the concept of the cellular system as a commercial venture. At the same time there were some "laboratory" tests carried out on equipment and operating techniques in Newark by AT&T, but no customers were involved in these tests.

The original AMPS system in Chicago consisted of ten cells covering about twenty-one-thousand square miles of the metropolitan area. In this system new techniques, hardware, and software were utilized and tested as a single concept. A central computer and switching system were used driven by SPC. Custom circuitry and microprocessors using Large-Scale Integrated (LSI), circuits were used throughout the system. All this technology was brought together for the first time to prove that the cellular concept was feasible.

AT&T, of course, planned to take the AMPS system nationwide after it had been proved so successful in Chicago. But the breakup of the telephone monopoly in January 1984 meant that the regional telephone companies would have to set up the cellular systems in their own areas. In addition, the FCC ruled that they would have to compete with other systems that would be licensed by the FCC to operate in the regional markets. These competing organizations were known as nonwireline companies, and the FCC set aside half of the available channels for assignment to them.

Before the divestiture in 1984, AT&T had planned to put the next system in the Washington-Baltimore area and incorporate improvements based on the experience gained in Chicago. The system was taken over by the regional telephone company, Bell Atlantic, and began operations in the middle of 1984 as Bell Atlantic Mobile Systems. Meanwhile, a nonwireline company, a consortium of communications and media organizations in the area, had gotten a cellular license and began operations. Table 2-1 shows the status of the installation and operation of the cellular systems in 90 areas around the country as of March 1986.

**TABLE 2-1 Status Report (Reprinted with permission of *Cellular Business*, May 1986, pp. 46-48.)**

| MSA #/Name | | System Operators | Status | # Cells |
|---|---|---|---|---|
| **1** | NEW YORK | W—Nynex Mobile | On line 6/15/84 | 48 |
| | | NW—Metro One | On line 4/5/86 | 24 |
| **2** | LOS ANGELES | W—PacTel Mobile Access | On line 6/13/84 | 29 |
| | | NW—LA Cellular Telephone | CPG 12/4/84 | 24 |
| **3** | CHICAGO | W—Ameritech Mobile | On line 10/13/83 | 44 |
| | | NW—Cellular One | On line 1/13/85 | 18 |
| **4** | PHILADELPHIA | W—Bell Atlantic Mobile | On line 7/12/84 | 32 |
| | | NW—Metrophone | On line 2/12/86 | 18 |
| **5** | DETROIT | W—Ameritech Mobile | On line 9/21/84 | 19 |
| | | NW—Cellular One | On line 7/30/85 | 15 |
| **6** | BOSTON | W—Nynex Mobile | On line 1/1/85 | 20 |
| | | NW—Cellular One | On line 1/1/85 | 10 |
| **7** | SAN FRANCISCO | W—GTE Mobilnet | On line 4/2/85 | 25 |
| | | NW—Cellular One | CPG 8/9/84 | 27 |
| **8** | WASHINGTON | W—Bell Atlantic Mobile | On line 4/2/84 | 38 |
| | | NW—Cellular One | On line 12/16/83 | 25 |
| **9** | DALLAS | W—Southwestern Bell Mobile | On line 7/31/84 | 41 |
| | | NW—Metroplex | On line 3/1/86 | 28 |
| **10** | HOUSTON | W—GTE Mobilnet | On line 9/28/84 | 8 |
| | | NW—Houston Cellular Telephone | CPG 12/27/84 | 29 |
| **11** | ST. LOUIS | W—Southwestern Bell Mobile | On line 7/16/84 | 17 |
| | | NW—CyberTel | On line 7/16/84 | 13 |
| **12** | MIAMI | W—BellSouth Mobility | On line 5/25/84 | 23 |
| | | NW—Florida Cellular Telephone | CPG 4/26/85 | 16 |
| **13** | PITTSBURGH | W—Bell Atlantic Mobile | On line 12/10/84 | 10 |
| | | NW—Cellular One | CPG 3/6/84 | 17 |
| **14** | BALTIMORE | W—Bell Atlantic Mobile | On line 4/2/84 | 38 |
| | | NW—Cellular One | On line 12/16/83 | 25 |
| **15** | MINNEAPOLIS | W—NewVector Communications | On line 6/6/84 | 13 |
| | | NW—MCI/Cellcom | On line 7/23/84 | 11 |
| **16** | CLEVELAND | W—GTE Mobilnet | On line 12/18/84 | 9 |
| | | NW—Cellular One | On line 5/31/85 | 7 |
| **17** | ATLANTA | W—BellSouth Mobility | On line 9/5/84 | 12 |
| | | NW—GenCom Cellular of Atlanta | CPG 1/18/85 | 10 |
| **18** | SAN DIEGO | W—PacTel Mobile Access | On line 8/15/85 | 8 |
| | | NW—GenCom | CPG 3/7/85 | |
| **19** | DENVER | W—NewVector Communications | On line 7/10/84 | 10 |
| | | NW—Cellular One | CPG 1/31/85 | 11 |
| **20** | SEATTLE | W—NewVector Communications | On line 7/12/84 | 13 |
| | | NW—Cellular One | On line 12/12/85 | 15 |
| **21** | MILWAUKEE | W—Ameritech Mobile | On line 8/1/84 | 9 |
| | | NW—Milwaukee Telephone Co. | On line 6/1/84 | 7 |
| **22** | TAMPA | W—GTE Mobilnet | On line 11/30/84 | 10 |
| | | NW—Bayfone | CPG 4/26/85 | |
| **23** | CINCINNATI | W—Ameritech Mobile | On line 11/5/84 | 13 |
| | | NW—Southern Ohio Telephone | CPG 1/9/85 | |
| **24** | KANSAS CITY | W—Southwestern Bell Mobile | On line 8/14/84 | 13 |
| | | NW—Cellular One | On line 2/14/86 | 12 |
| **25** | BUFFALO | W—Nynex Mobile | On line 4/16/84 | 7 |
| | | NW—Buffalo Telephone | On line 6/1/84 | 9 |

Key: W—wireline carrier. NW—non-wireline carrier. CPG—construction permit granted. Information available as of March 28, 1986.

| MSA #/Name | | System Operators | Status | # Cells |
|---|---|---|---|---|
| 26 | PHOENIX | W—NewVector Communications<br>NW—Metro Mobile CTS | On line 8/15/84<br>On line 3/1/86 | 9<br>10 |
| 27 | SAN JOSE | W—GTE Mobilnet<br>NW—Cellular One | On line 4/2/85<br>CPG 8/9/84 | 24<br>27 |
| 28 | INDIANAPOLIS | W—GTE Mobilnet<br>NW—Indianapolis Telephone Co. | On line 5/3/84<br>On line 2/3/84 | 5<br>9 |
| 29 | NEW ORLEANS | W—BellSouth Mobility<br>NW—Radiofone | On line 9/1/84<br>On line 9/6/85 | 5<br>5 |
| 30 | PORTLAND<br>OR | W—GTE Mobilnet<br>NW—Cellular One | On line 3/5/85<br>On line 7/12/85 | 5<br>7 |
| 31 | COLUMBUS<br>OH | W—Ameritech Mobile<br>NW—Cellular One | On line 5/30/85<br>CPG 1/28/85 | 5 |
| 32 | HARTFORD<br>CT | W—Southern New England Tel.<br>NW—Hartford Cellular Co. | On line 1/31/85<br>CPG 2/14/85 | 6 |
| 33 | SAN ANTONIO<br>TX | W—Southwestern Bell Mobile<br>NW—San Antonio Cellular Tel. | On line 1/28/85<br>CPG 1/30/85 | 12 |
| 34 | ROCHESTER<br>NY | W—Rochester Telephone<br>NW—Genesee Telephone Co. | On line 6/4/85<br>CPG 1/30/85 | 5<br>7 |
| 35 | SACRAMENTO<br>CA | W—PacTel Mobile Access<br>NW—Sacramento Cellular Tel. | On line 8/29/85<br>CPG 2/13/85 | 5<br>5 |
| 36 | MEMPHIS<br>TN | W—BellSouth Mobility<br>NW—Memphis Cellular Tel. | On line 5/1/85<br>CPG 2/13/85 | 5 |
| 37 | LOUISVILLE<br>KY | W—BellSouth Mobility<br>NW—Louisville Telephone | On line 1/3/85<br>On line 2/15/85 | 6<br>5 |
| 38 | PROVIDENCE<br>RI | W—Nynex Mobile<br>NW—Providence Cellular Tel. | On line 8/22/85<br>CPG 9/21/84 | 4<br>8 |
| 39 | SALT LAKE CITY<br>UT | W—NewVector Communications<br>NW—Salt Lake City Telephone | On line 1/29/85<br>CPG 3/6/85 | 6 |
| 40 | DAYTON<br>OH | W—Ameritech Mobile<br>NW—Cellular One | On line 5/31/85<br>On line 6/10/85 | 5 |
| 41 | BIRMINGHAM<br>AL | W—BellSouth Mobility<br>NW—Birmingham Cellular Tel. | On line 9/26/85<br>CPG 2/14/85 | 3 |
| 42 | BRIDGEPORT<br>CT | W—Southern New England Tel.<br>NW—Bridgeport Cellular Co. | On line 5/20/85<br>CPG 1/28/85 | 5 |
| 43 | NORFOLK<br>VA | W—Contel Cellular, Inc.<br>NW—Cellular One | On line 5/3/85<br>On line 11/1/85 | 4<br>5 |
| 44 | ALBANY<br>NY | W—Nynex Mobile<br>NW—Cellular System One | On line 6/25/85<br>CPG 9/4/84 | 4 |
| 45 | OKLAHOMA CITY<br>OK | W—Southwestern Bell Mobile<br>NW—Cellular One | On line 1/14/85<br>On line 1/10/86 | 9 |
| 46 | NASHVILLE<br>TN | W—BellSouth Mobility<br>NW—Nashville Cellular Telephone | On line 6/10/85<br>CPG 1/30/85 | 8 |
| 47 | GREENSBORO<br>NC | W—Centel<br>NW—Cellular One | On line 5/15/85<br>On line 12/27/85 | 8<br>9 |
| 48 | TOLEDO<br>OH | W—United TeleSpectrum<br>NW—Toledo Cellular Telephone | On line 7/25/85<br>CPG 12/8/83 | 9<br>7 |
| 49 | NEW HAVEN<br>CT | W—Southern New England Tel.<br>NW—New Haven Cellular Co. | On line 3/4/85<br>CPG 2/14/85 | 6 |
| 50 | HONOLULU<br>HI | W—GTE Mobilnet<br>NW—Honolulu Cellular Tel. | CPG 3/26/84<br>CPG 2/27/85 | 4<br>13 |
| 51 | JACKSONVILLE<br>FL | W—BellSouth Mobility<br>NW—Jacksonville Cellular Tel. | On line 6/12/85<br>CPG 2/21/85 | 6 |
| 52 | AKRON<br>OH | W—GTE Mobilnet<br>NW—Cellular One | On line 10/31/85<br>CPG 2/13/85 | 4 |

| MSA #/Name | | System Operators | Status | # Cells |
|---|---|---|---|---|
| 53 | SYRACUSE NY | W—Nynex Mobile<br>NW—Cellular One | On line 1/24/86<br>On line 12/31/85 | 3<br>3 |
| 54 | GARY IN | W—Ameritech Mobile<br>NW—Gary Cellular Telephone | On line 3/11/85<br>CPG 1/30/85 | 3<br>2 |
| 55 | WORCESTER MA | W—Nynex Mobile<br>NW—Worcester Cellular Tel. | On line 11/18/85<br>On line 11/18/85 | 5 |
| 56 | NORTHEAST PENNSYLVANIA | W—Commonwealth Telephone<br>NW—Northeast Pennsylvania Tel. | On line 7/2/85<br>On line 1/1/86 | 8<br>3 |
| 57 | TULSA OK | W—United States Cellular<br>NW—Tulsa Cellular Telephone Co. | On line 8/30/85<br>CPG 6/18/85 | 8<br>10 |
| 58 | ALLENTOWN PA | W—Bell Atlantic Mobile<br>NW—Cellular One | On line 3/18/85<br>On line 10/18/85 | 32<br>5 |
| 59 | RICHMOND VA | W—Contel Cellular, Inc.<br>NW—Cellular One | On line 5/10/85<br>CPG 2/4/85 | 5<br>7 |
| 60 | ORLANDO FL | W—BellSouth Mobility<br>NW—Orlando Cellular Tel. | On line 2/27/85<br>CPG 2/27/85 | 4 |
| 61 | CHARLOTTE NC | W—Alltel<br>NW—Metro Mobile | On line 4/15/85<br>On line 3/1/86 | 6<br>8 |
| 62 | NEW BRUNSWICK NJ | W—Nynex Mobile<br>NW—New Brunswick Cellular Tel. | On line 3/1/86<br>CPG 2/7/85 | 3 |
| 63 | SPRINGFIELD MA | W—Nynex Mobile<br>NW—Springfield Cellular Tel. | CPG 4/19/84<br>CPG 1/30/85 | |
| 64 | GRAND RAPIDS MI | W—GTE Mobilnet<br>NW—Grand Rapids Cellular Tel. | CPG 10/17/84<br>CPG 1/30/85 | 4<br>5 |
| 65 | OMAHA NE | W—Centel<br>NW—Omaha Cellular Telephone | On line 4/15/85<br>On line 12/23/85 | 4<br>3 |
| 66 | YOUNGSTOWN OH | W—United TeleSpectrum<br>NW—Youngstown Cellular Tel. | On line 9/19/85<br>On line 12/23/85 | 2<br>3 |
| 67 | GREENVILLE SC | W—GTE Mobilnet<br>NW—Metro Mobile | CPG 11/1/84<br>CPG 2/21/85 | |
| 68 | FLINT MI | W—Ameritech Mobile<br>NW—Flint Cellular Telephone | On line 7/12/85<br>On line 7/30/85 | 2<br>5 |
| 69 | WILMINGTON DE | W—Bell Atlantic Mobile<br>NW—Wilmington Cellular Tel. | On line 3/27/85<br>CPG 1/30/85 | 32<br>7 |
| 70 | LONG BRANCH NJ | W—Nynex Mobile<br>NW—Long Branch Cellular Tel. | CPG 9/26/84<br>CPG 1/30/85 | 3<br>4 |
| 71 | RALEIGH-DURHAM NC | W—United TeleSpectrum<br>NW—Cellular One | On line 11/11/85<br>On line 9/16/85 | 10<br>9 |
| 72 | W. PALM BEACH FL | W—BellSouth Mobility<br>NW—W. Palm Beach Cellular Tel. | On line 5/23/85<br>CPG 2/19/85 | 23 |
| 73 | OXNARD CA | W—PacTel Mobile Access<br>NW—Oxnard Cellular Telephone | On line 10/30/85<br>CPG 2/14/85 | 3 |
| 74 | FRESNO CA | W—Contel Cellular, Inc.<br>NW—Fresno Cellular Telephone | CPG 10/22/84<br>CPG 2/26/85 | 3 |
| 75 | AUSTIN TX | W—GTE Mobilnet<br>NW—Cellular One | On line 9/27/85<br>On line 12/27/85 | 5<br>8 |
| 76 | NEW BEDFORD MA | W—Nynex Mobile<br>NW—New Bedford Cellular Tel. | On line 12/9/85<br>CPG 2/13/85 | 2 |
| 77 | TUCSON AZ | W—NewVector Communications<br>NW—Metro Mobile | On line 8/6/85<br>On line 12/13/85 | 3<br>4 |
| 78 | LANSING MI | W—GTE Mobilnet<br>NW—Lansing Cellular Tel. | CPG 10/9/84<br>CPG 2/21/85 | 2<br>6 |
| 79 | KNOXVILLE TN | W—United States Cellular<br>NW—Knoxville Cellular Telephone | On line 7/23/85<br>CPG 2/7/85 | 7 |

| MSA #/Name | | System Operators | Status | # Cells |
|---|---|---|---|---|
| 80 | BATON ROUGE LA | W—BellSouth Mobility<br>NW—Baton Rouge Cellular Tel. | On line 7/2/85<br>CPG 1/30/85 | 3 |
| 81 | EL PASO TX | W—Contel Cellular, Inc.<br>NW—Metro Mobile | On line 2/25/85<br>CPG 1/28/85 | 2<br>4 |
| 82 | TACOMA WA | W—NewVector Communications<br>NW—Cellular One | On line 4/18/85<br>On line 12/12/85 | 3<br>17 |
| 83 | MOBILE AL | W—Contel Cellular, Inc.<br>NW—Bay Area Telephone Co. | On line 9/3/85<br>CPG 1/30/85 | 6 |
| 84 | HARRISBURG PA | W—United TeleSpectrum<br>NW—Harrisburg Cellular Tel. | On line 10/18/85<br>On line 9/18/85 | 4<br>4 |
| 85 | JOHNSON CITY TN | W—United TeleSpectrum<br>NW—Johnson City Cellular Tel. | On line 10/3/85<br>CPG 2/1/85 | 6 |
| 86 | ALBUQUERQUE NM | W—NewVector Communications<br>NW—Metro Mobile | On line 8/13/85<br>On line 11/1/85 | 2<br>3 |
| 87 | CANTON OH | W—GTE Mobilnet<br>NW—Canton Cellular Telephone | CPG 10/17/84<br>CPG 1/30/85 | 2 |
| 88 | CHATTANOOGA TN | W—BellSouth Mobility<br>NW—Chattanooga Cellular Tel. | On line 8/1/85<br>CPG 2/1/85 | 4 |
| 89 | WICHITA KS | W—Southwestern Bell Mobile<br>NW—Cellular One | On line 2/11/85<br>On line 1/24/86 | 4 |
| 90 | CHARLESTON SC | W—United TeleSpectrum<br>NW—Charleston Cellular Tel. | On line 9/11/85<br>CPG 1/28/85 | 5 |

While these efforts were going on to improve the mobile radiotelephone systems in automobiles, some work was being done about public communication in other forms of transportation. There were flurries of publicity from the railroads about installing telephone booths in some trains, but the problems encountered with the original mobile radiotelephones in automobiles were compounded on the trains. The trains moved out of the towns and out of range of the centrally located base stations quickly, and so calls had to be completed within a relatively short time.

Another problem existed in aircraft. The base station tended to be in range for a longer period of time, since the airplane was high in the air and, therefore, transmissions tended to cause interference with other systems.

The need for this type of instant communication in other than automobiles was not sufficient to warrant a great effort on the part of the telecommunications industry. After a few pictures and articles in home and hobby magazines, the instruments and the capability were removed.

Now, with the development of the cellular system, larger airplanes, and a traveling public educated to have a telephone always at hand, new efforts are being made in this field. These efforts are covered in more detail in Chapter 9.

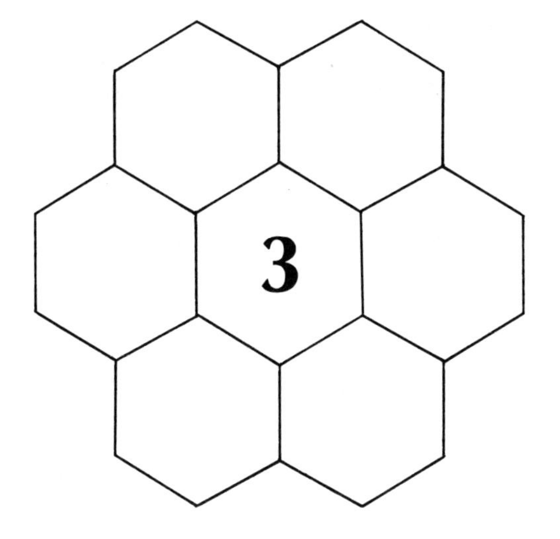


# The Technology in Detail

It may be that a lot of what follows in this chapter will be familiar to a number of readers. Explanations are necessary, however, to be sure that everyone gets off on the same foot. So, if this is old stuff to you, please bear with those to whom it may be a bit strange.

Before there is any discussion of the more exotic ways in which the human voice can be transmitted from one place to another, we should understand the basic principle of how the wireline telephone works. What happens when you speak into the mouthpiece of the instrument?

Look at a telephone handset the next time you use it. Most people at a young and curious age have at one time or another taken off the mouthpiece "to see where the voice goes" and found behind those holes in the mouthpiece a sort of capsule with a very thin metal diaphragm. Figure 3-1 shows a typical example.

When you speak, your vocal cords make waves in the air that correspond to the sounds you make. These waves go through the holes in the mouthpiece and make the thin diaphragm inside vibrate—move back and forth very quickly in accordance with the pitch of your voice. Tones that are low in pitch, such as bass voices, make the diaphragm vibrate at a slower rate than do tones that are high in pitch, such as soprano voices. Whether you speak in a loud voice or just whisper makes no difference to the *rate* of vibration. Loudness and softness

**Figure 3-1.** A transducer.

only increase the *amount* that the diaphragm moves. The rate, of course, is the number of times the diaphragm moves in a certain period of time, generally the number of times per second. The amount that the diaphragm moves is measured in thousandths and ten-thousandths of an inch but this measurement is rarely, if ever, used as a point of reference.

This diaphragm is mounted so that it presses against some carbon granules, through which a small electric current is passing—about 7 to 10 volts. One of the characteristics of these carbon granules is that the variations in pressure on them from the movements of the diaphragm, will vary the electrical resistance of the carbon to this voltage. This in turn varies the current flowing through the granules. Therefore the amount of current fluctuates in accordance with the vibrations caused by your voice. Those who remember Ohm's Law from elementary physics may recall how it applies here—if the voltage is fixed and the resistance is varied, the current will vary. This is exactly what takes place here.

The component is called a *transducer* which is what the engineers and scientists call a device that is actuated by power or energy from one system and supplies power or energy in any other form to another system.

Once your voice has been converted into these variations in electrical current—the name for this process is *modulation*—a lot of things can be done with it. The electrical current, which contains the information in your words, can be routed over any distance through any length of wire. It can also be amplified or attenuated as needed, changed in frequency so that it can be more easily handled in any subsequent operation, and broken up into smaller sections (called data bits) to be stored and manipulated. In addition, and of more importance to us, is the fact that this electrical current can also be converted into radio frequencies and transmitted anywhere without the need for the connecting wires of a telephone service.

Just generating a modulated signal is not enough, of course. The signal that is being sent over the wires or transmitted through the air must be rendered intelligible to the listener at the other end of the circuit. To accomplish this, the process just described works in reverse—the electrical signal ends up as a varying force in a magnet in the earpiece of the telephone. The magnetic force moves another diaphragm or vibrates it, and creates waves in the air that end up vibrating against your eardrum and being perceived as sound.

This is a greatly simplified explanation. Many refinements have been developed and added along the way to improve the efficiency and the quality of reproduction in the telephone system, but this is how it all begins whether you are still connected to the telephone system or are mobile with no tangible connection between you and the rest of the world. Figures 3-2 and 3-3 illustrate this concept.

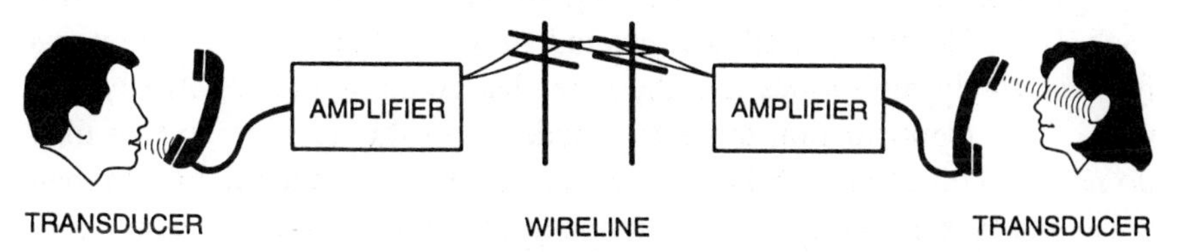


**Figure 3-2.** A basic wire communication system.

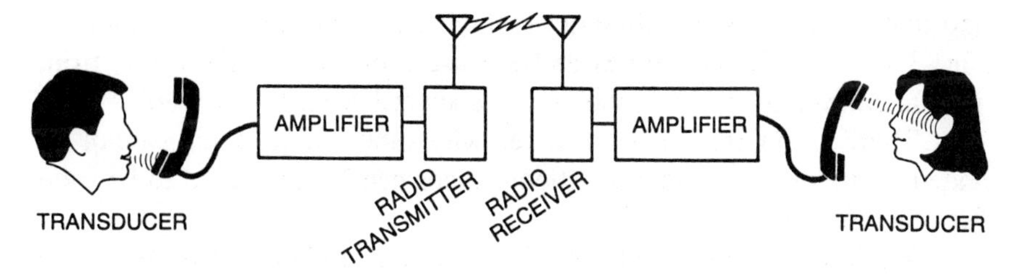


**Figure 3-3.** A basic wireless or radio communication system.

It is not enough, though, to get your message into shape so that you can send it along the wires of the telephone service. You have to make sure that the message gets to the right party. To do this there are two ways to signal the telephone system and tell it what number is wanted—rotary dialing and Touch-Tone. The first method generates a series of pulses in the line and the second method uses carefully selected tones as signals.

The rotary dialing system, along with the invention of the mechanical switch, put an end to the traditional operator who spent time at the switchboard pushing in and pulling out plugs to make and break the necessary connections. Rotary dialing instruments are still used by a lot of telephone subscribers, but because the information is transmitted by pulses—in reality, turning the power on and off—it is not as flexible as the Touch-Tone method. Rotary dialing is never used in cellular radio-telephone systems. In a landline or conventional system, by the way, the pulses generated by the subscriber with a rotary dialing instrument never go any further through the system than the local exchange. Here the pulses are converted to tones for use throughout the rest of the telephone system.

The Touch-Tone system is based on the use of a carefully selected series of tones. The term Touch-Tone is a registered trademark of the Bell Company and so must be capitalized whenever it is used. The generic term that is used in engineering texts is Dual Tone Multi-Frequency (DTMF).

The DTMF system is based on the use of two tones to identify each digit. The tones used are selected from two lists. One list contains four relatively high tones and the other list contains four relatively low tones. The identification tones are made up in this fashion so that they do not get confused with human speech. The tones generated by human vocal cords are made up of a number of components that are rarely equal in energy. The so-called pure tones generated by the DTMF method can be used to tell the system that a signal is being transmitted and that someone is not singing or playing an instrument. The table in Fig. 3-4 shows the two tones that are used for each of the digits from zero to nine and Fig. 3-5 shows the standard DTMF keyboard.

There is no reason, of course, why other frequencies cannot be added to the matrix so that additional tones can be generated to be used for other functions.

We have brushed over the DTMF system rather lightly, but there is no need to go into great detail regarding the stringent specifications

| Low band frequencies (in cycles per second) | High band frequencies (in cycles per second) 1209 | 1336 | 1447 | 1633 |
|---|---|---|---|---|
| 697 | 1 | ABC 2 | DEF 3 | |
| 770 | GHI 4 | JKL 5 | MNO 6 | |
| 852 | PRS 7 | TUV 8 | WXY 9 | |
| 941 | * | 0 | # | |

**Figure 3-4.** The Dual Tone Multi-Frequency (DTMF) table of tones.

**Figure 3-5.** A DTMF keyboard.

that cover the generation and use of the tones and their transmission. This factor has become increasingly important as the telephone system is used more and more for data transmission. For our look at the cellular mobile radiotelephone systems, though, it is sufficient to know generally how the DTMF system works and that it is used in cellular systems.

There is one more term that should be discussed since you will run across it rather frequently in the literature and in equipment and system specifications. This term is the decibel (db or dB). This term is used to describe the ratio between two power measurements. It was developed by Alexander Graham Bell during his work on measurements of hearing of the human ear. His test subjects said that a test sound was "twice as loud" when the power producing that sound was increased by a factor of ten over the power producing the previous sound. In mathematics, this is called a logarithmic relationship and is expressed by the formula:

$$\text{dB} = 10 \log \frac{P_2}{P_1}$$

This relationship was named the Bel after the inventor, but it was found that the quantities that resulted from the use of the original formula were too large to be handled conveniently. Therefore the term decibel, a tenth of a Bel, became the standard unit of measurement.

If you work through an example or two using this formula, you will find that the often seen phrase ". . . a 3-dB gain . . ." means that the power has been doubled. Four times the power gives a 6-dB gain and one-hundred times the power means a 20-dB gain. You will see this measurement used a lot in antenna specifications.

But enough about generalities. Let's talk more specifically about radiotelephone systems. No matter what the ads say, all the cellular mobile radiotelephone systems are essentially the same. The various pieces of equipment that make up the system might have different names because they are made by different manufacturers, but they all perform the same function. In some cases it may be the same equipment with just a different nameplate or logo on it.

In the vehicle is a handset, the transducer we talked about earlier, that the subscriber uses in the same manner as a regular telephone. The handset is perhaps a little more elaborate than an ordinary telephone, and it certainly has a lot more functions, as you will soon see, but it is still a handset. The user talks into it or, as in the *hands-off* mode of operation, talks at it. Figure 3-6 shows a typical instrument.

The handset is connected to the transceiver which is a combination of an FM radio transmitter, receiver, and logic unit. The logic unit controls the various functions of the transceiver and generates the housekeeping data for the cell site and the MTSO. These are packaged in a small cabinet that is mounted in a corner of the car trunk or perhaps behind the seat in a pickup. In the portable units, everything is mounted in the briefcase. The user need never do anything to the transceiver and, in fact, there are no controls available to adjust.

The transceiver is connected to an antenna, which is usually placed on the lid of the trunk or on the roof of the car, or it can be cemented to the rear window. The antenna is about a foot long with a little spiral of four to six turns in it that distinguishes it from all the other types of vehicle antennas that are in use such as those used by radio amateurs, Citizen Band (CB) users, taxi drivers, police, and others using various mobile land telephone services.

Now we come to the concept that named the system—the cell. This is perhaps the heart of the system and one of the main reasons for its success.

The cell is an area generally shown in a six-sided honeycomb

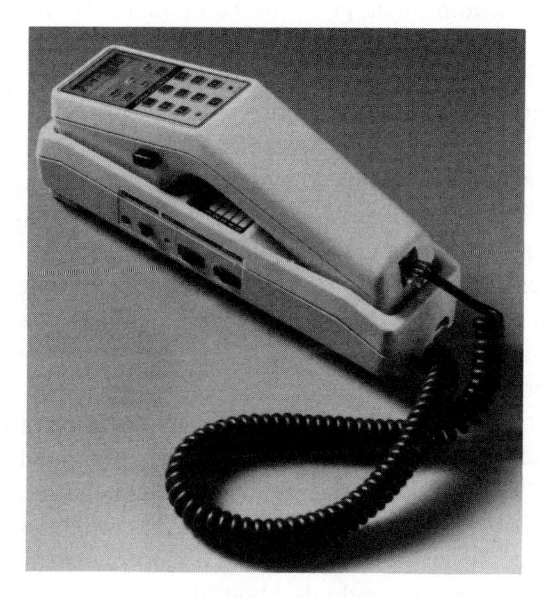

**Figure 3-6.** A cellular transducer. (Courtesy of Harris Corp.)

shape because this is the simplest way to illustrate how the system works. Figure 1-1 shows the idea. Depending on the terrain and/or the degree of urbanization of the location, the cell will cover an area about five times in diameter. Some cells are much smaller because of the amount of traffic or the character of the terrain, and some cells are larger. For example, the Washington-Baltimore system covers both downtown areas and the suburbs with sixteen cells.

The approximate boundaries of the cells are established by the location of the radios and antennas that serve as the link to the subscribers. In the industry this is known as the cell site. Again, Fig. 1-1 illustrates an idealized concept of the components of a complete system.

At the next level is the control center for the entire system. The name of the control center depends on who built it. The original name chosen by Bell, and the one that has been adopted by a lot of system operators, is the Mobile Telephone Switching Office, or MTSO. Motorola calls their control center the Electronic Mobile Exchange (EMX). Ericsson, a Scandinavian firm, calls it AXE; NEC, the Japanese telecommunications giant, calls it NEAX. Some of the engineering firms who assemble a complete cellular system with equipment from a number of manufacturers give the control center a different name. In any case, this vitally important piece of equipment, no matter what the name, is built around the technique of digital switching and computer control. The real secret is the software—the computer program that runs the system.

It might be well to remember before we get enmeshed in a lot of electronic details that the control center performs the same functions as the old-time manually operated switchboard complete with operators. A piece of electronic gear called the controller does it all without any human intervention—the routing of the call, the timing and billing, evaluating any trouble on the line—all of it. It does everything faster and, without downgrading the work of the early operators, does it all with a lot more efficiency.

Most of the people in the industry, when they are not talking about or trying to sell one particular manufacturer's item, will refer to the controller as "a switch" or "the switch" because that is really what it is. This is probably a carry-over from the telephone industry where so many of the people in the cellular system came from.

Let's go back a little and talk about the type of radio transmission that is used in the cellular system. It is a narrow-band, frequency modulated (FM) system operating in the 800-MHz band. You are probably familiar with this type of broadcasting mainly because it is used for the higher quality transmission on our commercial and public broadcast stations. Its static-free, high fidelity is important for cellular use too. However, more important is the so-called capture effect. This is a characteristic of FM in which a receiver detecting two separate FM transmissions on the same frequency will ignore the weaker signal and process the stronger without any real loss in the signal quality. In the cellular systems we are looking at, this means that if the same channel is being used by two different cells for different subscribers, the receiver will not bother with any weaker signal which is probably originating from a distant cell and being used by a different subscriber.

To keep this sort of interference to a minimum, the radio frequency (RF) power used to transmit the signal between the mobile subscriber and the cell site being used is being constantly monitored, measured, and adjusted to the lowest level of power that will give effective communication. This monitoring, along with other housekeeping and data-collecting functions, is being done at all times. However, it is done on another radio channel rather than on the voice channel. This is called a *setup* channel and the data being transmitted over this channel is binary in form—there is never any voice transmission over this channel.

These, then, are the components that make up the cellular mobile radiotelephone system. Each piece of equipment is described and illustrated in detail in later chapters.

Now that the individual components that make up a cellular system have been described, let's go slowly through what happens when a call is initiated or received. The description that follows will apply to any of the cellular units—the mobiles mounted in a vehicle, the handheld portables, the units packaged in a briefcase. All these units use similar circuits and computer programs and are only packaged differently.

When an instrument is turned on or *powered up*, it enters what is called the *idle state*. In this condition it is only doing the electronic housekeeping routines that are necessary to prepare it to receive or make calls. The logic unit in the transceiver directs the receiver portion to scan a band of frequencies, or channels (called paging or setup channels), to find which one of them has the strongest signal. This generally, but not always, turns out to be the one from the closest cell site.

Having found the strongest signal, the receiver passes the information on to the logic unit that directs the receiver to lock onto that frequency and monitor it continuously or until it receives instructions to switch to another channel. At intervals the receiver will rescan the setup channels to see if a different channel will give a better quality signal. The most obvious reason for doing this, of course, would be that the station has moved into a location where it can receive a better signal from another cell site. If this is the case, the receiver is directed to switch to that channel.

So far, remember, the subscriber has merely turned on the unit—nothing more. To place a call to a mobile station from a fixed location, the caller dials the mobile station's number. This number will be the usual seven digits in length but with the first three numbers coded so that the local telephone system will route the call to the cellular switch or controller. This is exactly the same procedure used to route a call to the correct central office over the landlines of the PTSN.

When these digits arrive at the MTSO, they are checked through the memory bank in the switch computer and, if recognized are matched up with a separate identification number and sent out to all the cell sites together with instructions to transmit this data on all the paging channels.

The called unit recognizes its Identification Number (ID) among all the data transmitted on the paging channels as it searches them. On another setup channel, the unit transmits its ID to the cell site, which in turn sends it to the MTSO. The MTSO then assigns a voice channel that is not in use to the called unit that switches automatically to that voice channel and sends a signal to the cell site confirming that it has done so.

The cell site then notifies the switch that the connection has been made and that the number has been reached. The switch commands the cell site to send a signal over the setup channel to notify the mobile unit that it is being called. When the mobile station answers the call, it goes *off hook* and the cell site acknowledges that the signalling tone has stopped. The cell site signals this fact to the switch, which then connects the call to the mobile unit. The controller also notifies the billing computer to start billing the subscriber who pays the bill whether the call is outgoing or incoming.

The process is a little different for calls originated by the mobile station. The subscriber punches in the desired number before the handset is picked up. The number is stored in a register until the subscriber is ready to place the call. Then the "send" button is depressed.

This tells the mobile transceiver unit to transmit a digital message to the cell site giving the station's ID number, the called number, and a request for a voice channel. The cell site relays this information and the request to the switch where a vacant voice channel is assigned to the cell.

The mobile unit receives this new assignment via the cell site and the logic unit automatically switches the transceiver to the assigned channel. The switch completes the call through the landline system for handling by the telephone company.

If the mobile station gets a busy signal, the handset is replaced as in a normal telephone call. The desired number is still stored in the little memory in the register in the handset, and so the number can be re-called at any time by simply pressing the "send" button for another attempt. The whole process is then repeated. When the connection is made the computer begins to time the call for billing purposes.

One of the better ways of presenting what appears to be a rather complicated procedure is by using a table such as that shown in Figs. 3-7 and 3-8. These tables were taken from the OKI telecommunications organization publication titled *Basics of the Cellular Telephone System*. Words and terms that are unfamiliar are found in the glossary.

What happens when the mobile or portable station begins to move out of range of the signal from the cell site while in the middle of a call? This is not an unusual situation at all with normal cells 1½ to 5 miles across and a mobile station moving from one side of town to another.

To begin, the strength of the signal on the existing channel between the cell site and the mobile station is being constantly monitored over the data channels and compared with a standard level of strength. When the signal gets weak and goes below the standard, the MTSO transfers the call to a vacant channel that will give a better signal either in the same cell site or in an adjacent cell site. At the same time, the cell site is directed to tell the mobile unit to switch to the newly assigned channel. The mobile unit does this automatically and the MTSO checks to see that the switch has taken place. This procedure, called a *hand-off*, takes place in less than 0.2 sec and is accomplished without either party being aware of what has taken place.

When the call has been completed and the mobile subscriber

**MOBILE CALL**

| | MTSO | CELL SITE | MOBILE UNIT |
|---|---|---|---|
| 1. | | Transmits overhead information on control channel | |
| 2. | | | Scars and locks-on best control channel |
| 3. | | | User pre-originates call user presses "SEND" key |
| 4. | | | Scans and locks-on best control channel |
| 5. | | | Seizes control channel |
| 6. | | | Sends service request |
| 7. | | Reformats service request and relays request to MTSO | |
| 8. | Selects voice channel and sends channel designation to cell | | |
| 9. | | Formats a channel designation message and sends to mobile via control channel | |
| 10. | | Prepares voice channel and transmits SAT | |
| 11. | | | Tunes to assigned voice channel |
| 12. | | | Detects and verifies then retransmits SAT |
| 13. | | Detects and verifies SAT | |
| 14. | | Puts off-hook on trunk to MTSO | |
| 15. | Detects off-hook | | |
| 16. | Complete call through network | | |

**Figure 3-7.** The steps in a cellular telephone call transmission. (Courtesy of OKI Telecom, Cellular Telephone Division.)

**MOBILE ANSWER**

| | MTSO | CELL SITE | MOBILE UNIT |
|---|---|---|---|
| 1. | | Transmits overhead information | |
| 2. | | | Scans and locks-on best control channel |
| 3. | Receives incoming calls | | |

**Figure 3-8.** The steps in a cellular telephone call reception. (Courtesy of OKI Telecom, Cellular Telephone Division.)

| | MTSO | CELL SITE | MOBILE UNIT |
|---|---|---|---|
| 4. | Sends paging message to cell sites | | |
| 5. | | Reformats paging message and sends to mobile unit via control channel | |
| 6. | | | Detects page |
| 7. | | | Scans and locks-on best control channel |
| 8. | | | Seizes control channel |
| 9. | | | Sends service request |
| 10. | | Reformats service request and relays request to MTSO | |
| 11. | Selects voice channel and sends channel designation to cell | | |
| 12. | | Formats a channel designation message and sends to mobile via control channel | |
| 13. | | Prepares voice channel and transmits SAT | |
| 14. | | | Tunes to assigned voice channel |
| 15. | | | Detects and verifies then retransmits SAT |
| 16. | | Detects and verifies SAT | |
| 17. | | Puts off-hook on trunk to MTSO | |
| 18. | Detects off-hook | | |
| 19. | Sends alert order | | |
| 20. | | Reformats alert order and relays to mobile via blank-and-burst on voice channel | |
| 21. | | | Generates ringing and sends 10-kilz tone to cell |
| 22. | | Detects 10-kilz tone | |
| 23. | | Puts on-hook on trunk | |
| 24. | Detects on-hook | | |
| 25. | Provides audible ring to calling party | | |
| 26. | | | User answers call |
| 27. | | | Discontinues 10-kilz tone |
| 28. | | Detects absence of 10-kilz tone | |
| 29. | | Puts off-hook on trunk | |
| 30. | Detects off-hook | | |
| 31. | Removes audible ring and completes connection | | |

**Figure 3-8 cont.**

hangs up (goes *on hook*), the process is essentially reversed. The transceiver in the mobile installation gives up the voice channel that had been in use and resumes scanning the data channels for another call. The cell site signals the MTSO that the voice channel is no longer in use and is free for other users and the MTSO breaks the connection to the PTSN that had been in use. The total time of the call is recorded against the subscriber and added to the bill that is being compiled by the computer that handles that task.

There is little doubt that the development of the cellular mobile radiotelephone system would not have been possible without the evolution of the minicomputer. This wonder of the electronic age has become such an integral part of our lives that we have come to accept it and tend to forget how we got along without it.

The use of this capability has become so pervasive that the lines of definition have become somewhat blurred. Just what is a computer? Or, more correctly, what is a minicomputer? Can you call the little black box in your automobile that monitors the temperature of the coolant in your radiator a computer? This devise measures the temperature of the coolant and when the temperature reaches some set limit, it turns on the cooling fan. When the temperature goes down again, it turns the fan off. This device certainly monitors and controls one of the many variables in the operation of your automobile.

How about the controls of your microwave oven? The device will turn the oven on and off at predetermined times as well as monitor and control the cooking of the roast by means of a probe.

These devices are called *dedicated units*. This means that they are designed to perform one particular task and they cannot be reprogrammed to do something else. The instructions, or program, that the unit is to follow is designed into the unit, unlike the personal computer sitting at home in the study or the living room. The personal computer can be programmed to do many things and the shelves of the computer stores are full of different programs for it. However, if one of these personal computers were made part of a system and used to do only one thing and nothing else, it would then also be considered a dedicated computer.

The cellular system is a mix of dedicated computers and general-purpose computers. The mobile or portable transceivers contain some integrated circuits (IC) whose instructions are "burned in." This will be described in Chapter 6. Another role for a dedicated computer is described in Chapter 5 in which the cell site and its equipment is discussed.

The job of the computer at the MTSO is to control the digital switching system and do all the other tasks such as billing, traffic analysis, and monitoring the operation of all the cell sites. In some systems the computer will be a general-purpose unit that has been programmed to operate the digital switch as well as do all of the technical and bookkeeping tasks that we have mentioned. In other systems the computer capability is an integral part of the switch. The switch/computer combination is an outgrowth of the digital switching technology developed by the telephone industry. Other stand-alone computers, the number depending upon the size of the system, are programmed and used for other tasks, such as testing and monitoring.

You will recall that we talked a little about the characteristics of the technology that permitted the reuse of the frequencies in an area. A rule of thumb in the industry states that a channel can be reused again in every seven cells. The application of this rule will vary tremendously, of course, since there are so many factors that will affect the propagation of the radio transmissions. The particular terrain, the degree and density of urban growth, the expected number of subscribers—all these things will determine the size of the cell. However, a figure is needed for planning purposes and experience has shown the industry that this rule of thumb is a reasonable one.

In an area like Washington-Baltimore, where the initial plan called for 16 cells, engineers could use each frequency 2.28 times—16 cells divided by 7. Since there are a total of 333 channels or frequencies available, multiplying the number of channels by the number of times each channel can be reused means that there will be about 760 channels available—333 times 2.28—throughout the entire system.

Further, it has been determined that each channel will handle between 20 and 25 subscribers without an unacceptable number of busy signals—not everyone will be trying to make a call at the same time. Multiplying all these figures together—760 channels times 20 subscribers per channel—gives a very rough potential capacity of fifteen-thousand-two-hundred subscribers.

This figure was reached using greatly oversimplified data. The cellular industry is still too new to have generated a lot of experience that can be used in system design. The foregoing exercise shows how available data is used.

Since the system operator cannot add channels to the system—each system is limited to 333 channels—a technique called channel splitting has been devised. This technique is demonstrated in the examples illustrated in Fig. 3-9. The area is covered by seven cells, which are shown in the conventional shape and with an antenna located in the middle of each cell. This situation never happens in real life—it is never anywhere as neat and tidy as this—but this example illustrates the technique involved.

By changing the antenna design in one cell so that the RF energy radiation is concentrated in an arc of 120° rather than the complete 360° circle used formerly, the number of cells can be increased (see Fig. 3-10). Using this technique the same cell site can be used. Only the antennas have to be changed, additional equipment installed, and computers instructed to handle the additional load. This can be carried on again and the original seven-cell system could end up with twenty-one cells, as illustrated in Fig. 3-11.

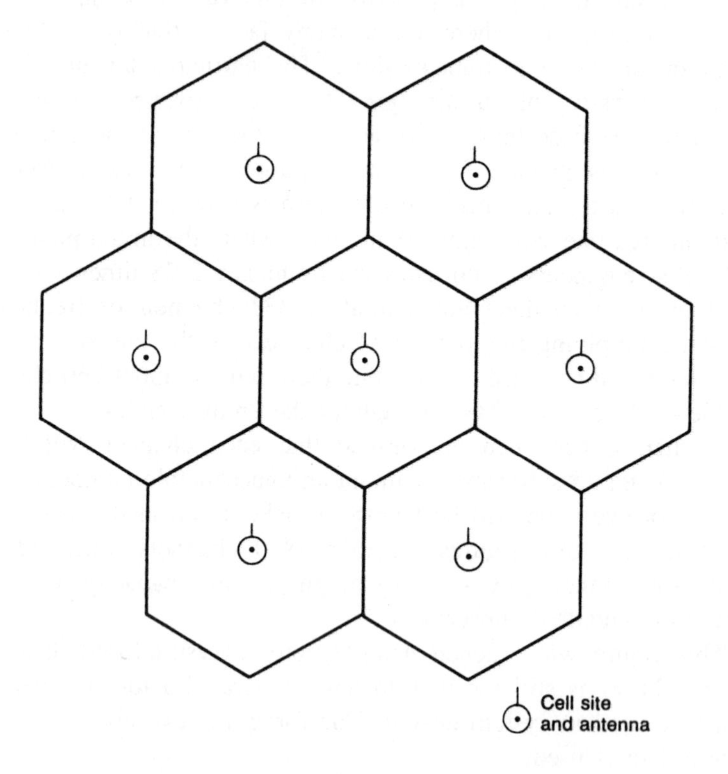


**Figure 3-9.** An idealized seven-cell configuration.

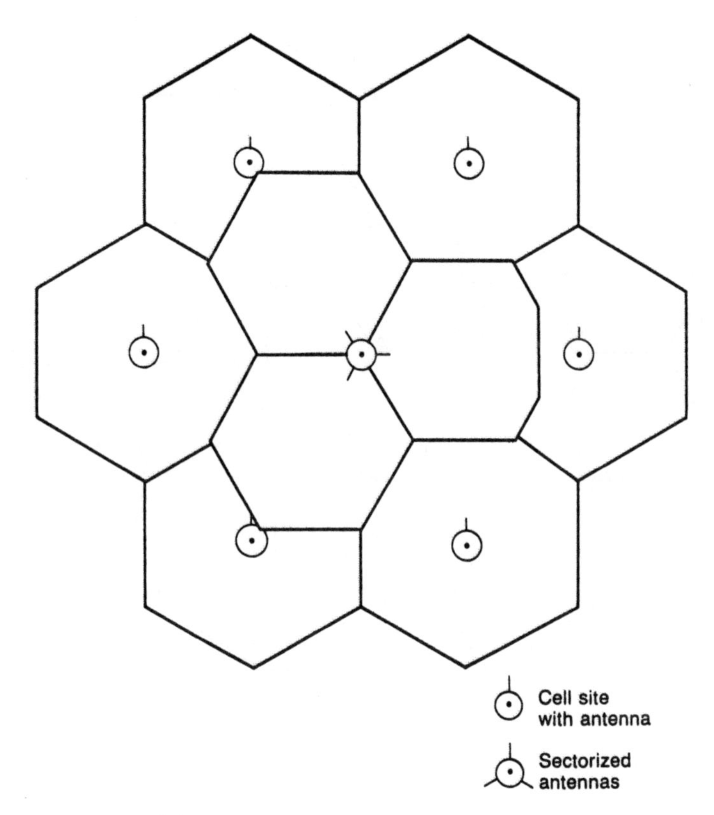


**Figure 3-10.** An example of cell splitting.

None of this cell splitting can be done without extensive testing and engineering studies. One of the tasks of the computer at the MTSO is traffic analysis, enabling the system manager to sense subscriber needs and changes well in advance so that tests and studies can be carried out. While the cellular equipment is not "far out" in terms of design and construction, it cannot presently be purchased off the shelf at the neighborhood electronics shop. It takes time to analyze data, make necessary engineering studies, and design and install equipment to give the increase in subscriber capacity.

All that has been described so far has been designed and installed to keep the system working efficiently and the customer happy. When the subscriber drives outside the home CGSA and the coverage by that one particular system, it may be another story. The term *roamer* is used by the industry to describe a subscriber who roams from the area covered by one system to the area covered by another system and wants

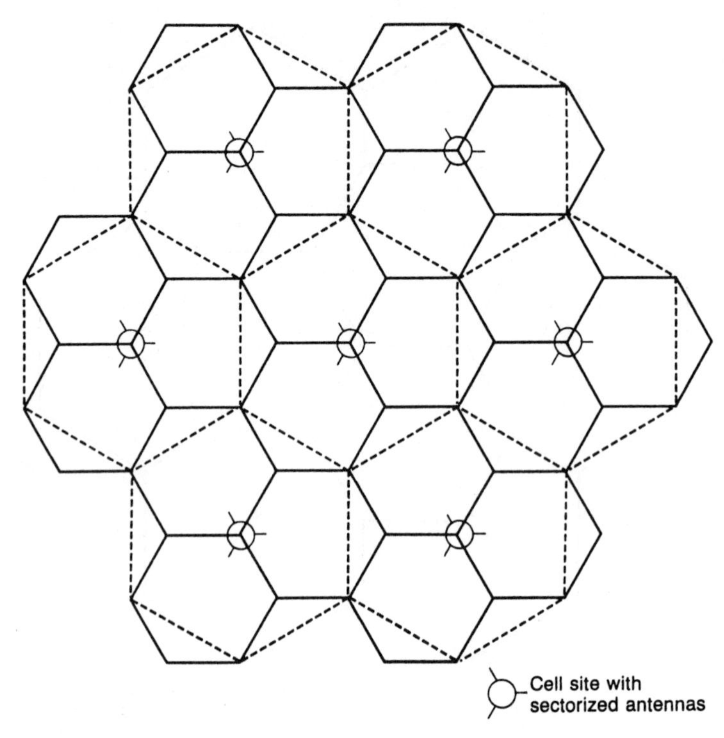


**Figure 3-11.** Idealized expansion to a twenty-one cell system.

to make a call. A lot has been written about this particular problem. In the more densely populated areas (the Washington-Boston corridor is a typical example), a number of systems cover the area so that a subscriber to one of the Washington area cellular systems, for example, can pick up the cellular handset anywhere along the New Jersey or Connecticut turnpikes and make a connection through the system with a cell site in that area.

However, unless the subscriber's home system (in this case one of the Washington systems) has reached a roaming agreement with the system covering the area through which the customer is driving, the "no service" indicator will light up on the instrument and the call will not be completed. The customer will have to stop at one of the service areas and make the call from a wireline system call box just like anyone else. Avoiding this was probably one of the main reasons the cellular radiotelephone was installed in the car.

With conventional wireline telephone systems, the problem of making a call from one locality to another is taken care of by AT&T.

Each of the seven systems that cover the country after the breakup of AT&T, are interconnected by the long distance lines of that company. This was one feature of the old monopoly that remained after the breakup of AT&T.

This same feature does not yet exist with the cellular systems—there is no overall organization covering all the cellular systems that are up and running throughout the country.

This problem is recognized by the industry and through a Roaming Committee of the trade association, the Cellular Telecommunications Industry Association (CTIA) is attempting to find a solution. Some kind of industry-financed central sorting office would be one answer. Bills for calls made by roamers could be sent, sorted, and distributed much like the checks that are sent all over the country by the banking industry.

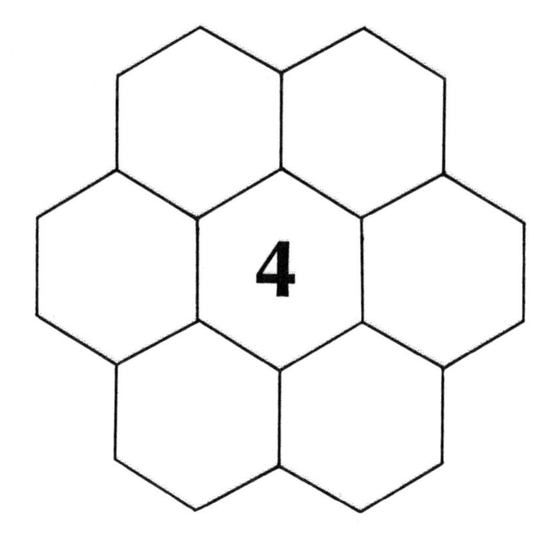


# The Heart of the Matter

The equipment that we are going to talk about now is really the guts of the cellular mobile radiotelephone system. It consists of two main components: the switch and the computer. These components can be separate, stand-alone items, or they can be combined into one unit designed specifically for this application. Figure 4-1 shows how Motorola illustrates their switch/computer combination that has been designed for cellular systems. As the number of subscribers increases and the capacity of the existing switch is reached, additional switches can be connected without changing the original configuration.

These components are the ones that separate the cellular system from all other mobile radiotelephone systems. It is these components, together with the instructions that run the computer (the software), that make possible the unique features of the cellular radiotelephones. The hand-off technique as the subscriber drives from one local area to another, the frequency reuse, the roaming from one system to another system many miles away—none of these features is possible with the other mobile radiotelephone systems. However, these features can be accomplished easily with cellular radiotelephones, in addition to the necessary administrative functions such as billing, traffic analysis, and gathering statistical data about the system. The computer also contains programs for routine testing and maintenance functions.

The computer maintains a file of subscriber's numbers or IDs that

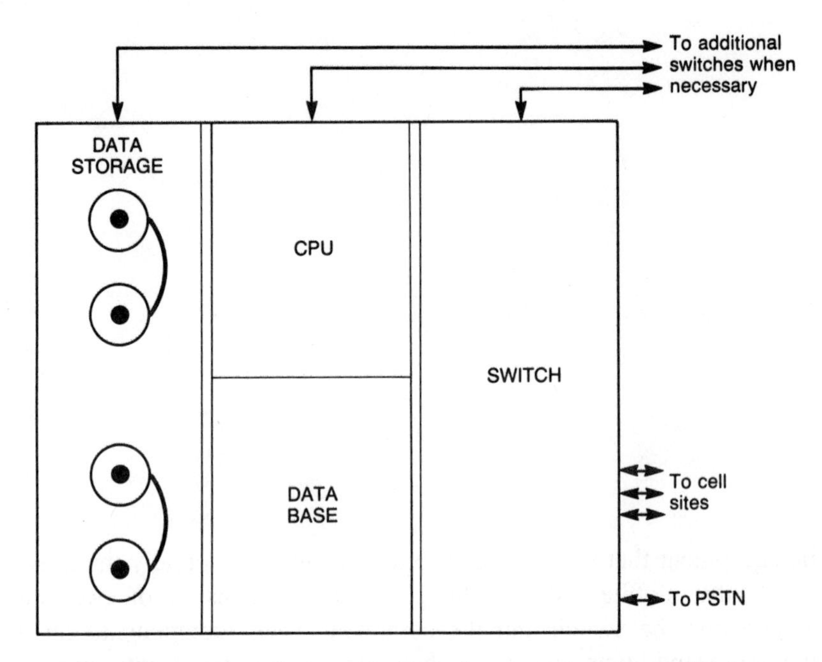


**Figure 4-1.** Block diagram of a typical computer/switch combination.

are not valid for some reason or another. The number may be assigned to an instrument that was stolen or to a subscriber with an outstanding bill. The caller may have been a roamer from an area where the system did not have a roaming agreement. In any case, after a review of the file in the memory, if the computer turns up a "forbidden" ID, the call will not go any further.

Last, this set of components serves as the interface between the cellular system and the PTSN, the public switched telephone network or, as you probably know it, the local telephone company.

Now a word or two about each of the components. The system switch—a typical one is shown in Fig. 4-2—performs just about the same functions as the manual switchboard pictured in Fig. 2-3. Rows of operators sat in front of these switchboards connecting and disconnecting the subscribers by plugging and unplugging the cords by hand.

The switch shown in Fig. 4-2 is intended strictly for cellular use. Other systems are designed to use a switch from the telephone industry. They drive or control it with a general purpose computer such as the Digital Equipment Company VAX (Fig. 4-3). The computer then becomes known as a dedicated computer, programmed and used for only one purpose as discussed in Chapter 3.

**Figure 4-2.** A typical digital switch, the Motorola Electronic Mobile Exchange (EMX). (Courtesy of Motorola, Inc.)

**Figure 4-3.** A general purpose computer, Digital Equipment Corporation's VAX, that can be programmed to control a switch. (Courtesy of Digital Equipment Corp.)

In its simplest form, any switch in an electronic circuit is only a method of connecting and disconnecting two points in a circuit. A simple switch such as a household light switch is depicted in Fig. 4-4. It connects inlet 1 to outlet A controlling the flow of current.

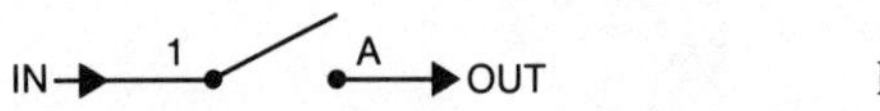


**Figure 4-4.** A simple switch.

The beginnings of a switching system (a matrix of switches) is shown in Fig. 4-5. In this type of schematic diagram, the lines are not connected where they cross. This is only one way to show a number of inlets and outlets in a circuit. This is an example of a four by four format: four lines going in and four lines coming out.

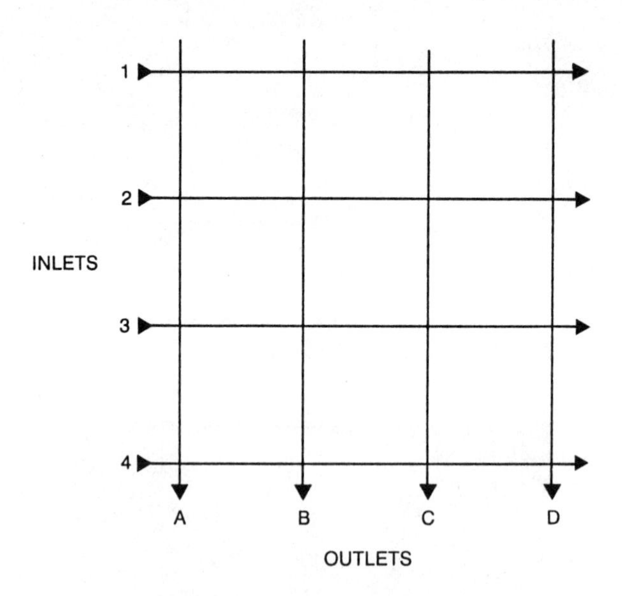


**Figure 4-5.** A 4 × 4 matrix.

In Fig. 4-6 this same four by four matrix has had switches added to it to show how any one of the input lines can be connected to any one of the outlet lines. The switches have been numbered S1 to S16 so that a path can be easily traced through the switch. This is the principle of the telecommunications switch.

The switches that will work in an application like this can be of any type—from a simple manually operated, flip-on, flip-off toggle switch to the computer-driven switches used in the cellular systems.

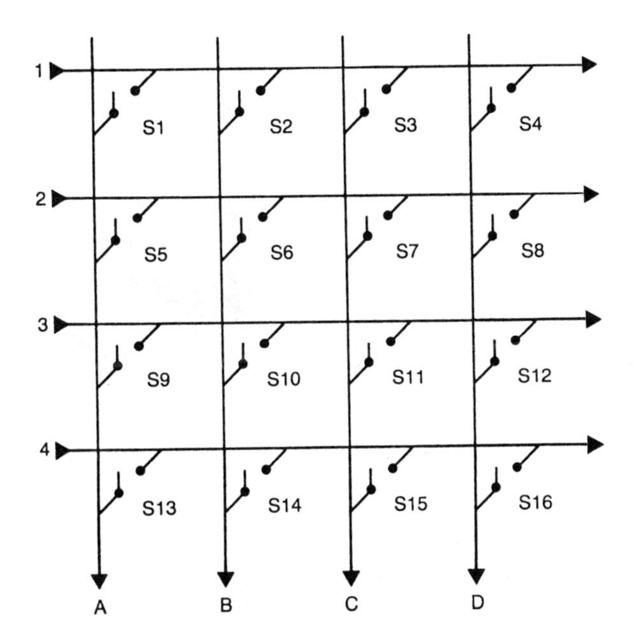


**Figure 4-6.** A 4 × 4 switching matrix.

The much-used phrase *digital switch* has become commonplace in the industry even though some of the information that is being switched or is used to control the actual switching function may not be in digital form.

A fully detailed technical explanation of the difference between the digital and the analog forms of information will not be discussed here, but it is enough to say that an analog signal is a continuous signal like the human voice. A digitized version of a signal would be one which was chopped up into a series of on-off signals and transmitted in that form. Such a simple version may not satisfy a purist but it seems to be enough to explain what takes place in a cellular telephone system.

As a quick example of how it works, let's follow the path for a circuit in which we want to connect input line 2 to outlet C in Fig. 4-7. This can be done by following input line 2 to the point where it intersects outlet line C and then closing switch S7 as shown. Another example: Input 4 can be connected to outlet B by closing switch S14 (Fig. 4-8). In a real situation, the switches are opened and closed electronically, of course, on instructions from the computer. There are integrated circuits, ICs, which are groups of tiny transistors that are designed to do the job of switching. Transistors use very little power and operate infinitely faster than any mechanical switch.

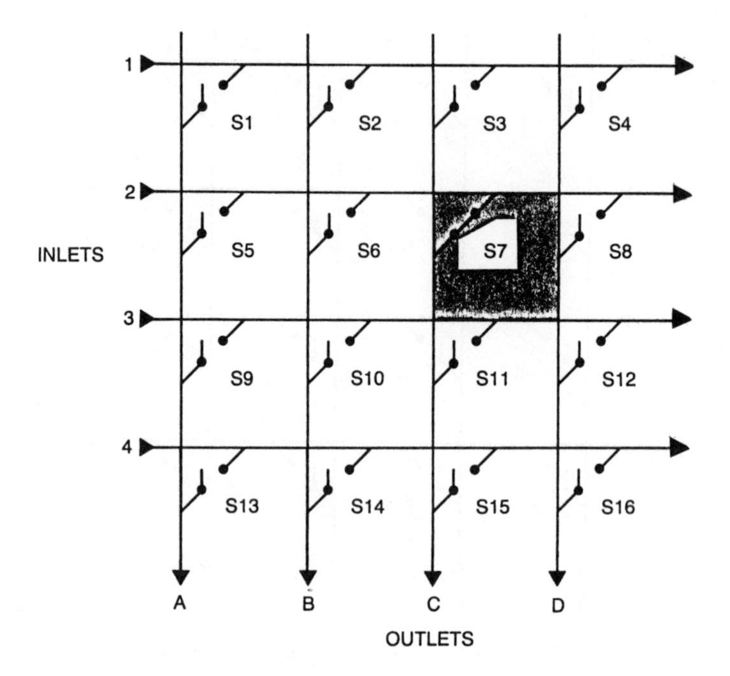


**Figure 4-7.** Connecting inlet 2 with outlet C.

The generic term for the task of switching is stored program control (SPC). All this technology, the switch and the computer operating under SPC, comes mostly from the telephone industry but has been modified to handle the radio frequencies of the cellular system rather than the audio frequencies of the telephone. Additional computer capacity and a larger program is needed since the computer has more duties when used in a cellular system.

The theory of digital switch design and the associated computer programming have been the subject of much study in the telecommunications field. Many technical books and papers have been written on this esoteric topic. But you can always come back down to earth by remembering that the switch/computer combination is just an automated old-time switchboard. Of course, it has taken millions of research hours and the same number of dollars to reach this state. When you feel that you are getting out of your depth, just remember the operator answering the telephone with the traditional "Number, please" when you pick up the receiver or go *off-hook*, as they say in the business.

As we have said, the MTSO comes from the telephone industry and so most of the equipment in the cellular systems is the result of a

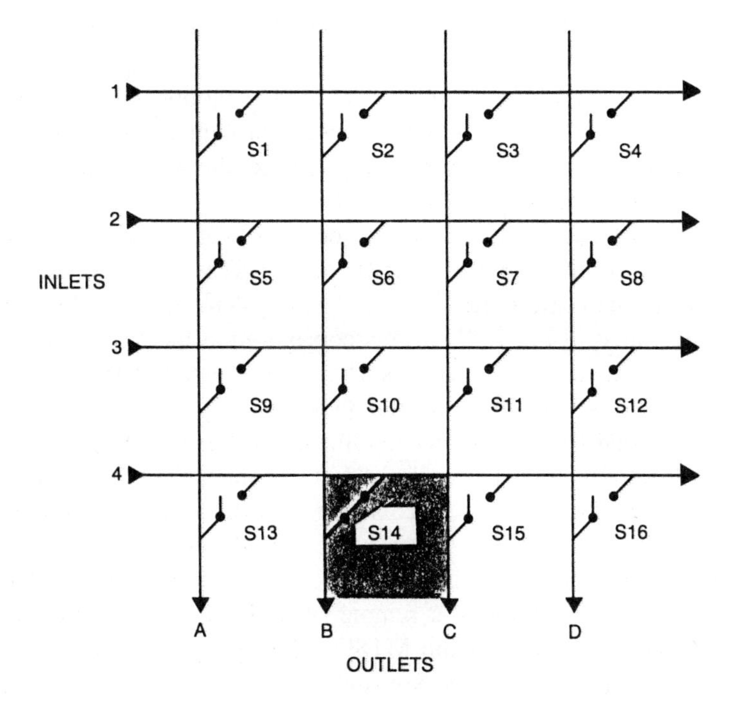


**Figure 4-8.** Connecting inlet 4 with outlet B.

joint effort of firms experienced in the telephone and radiocommunications businesses—a merging of techniques and experience that has paid off for both firms.

This combination of switch and computer, together with the interface with the landlines of the PTSN, and the control and administrative functions associated with the system, has been given a number of names or acronyms. Probably the most descriptive and most accurate name that correctly lists all the functions is the one that we have been using: mobile telephone switching office. AT&T, in the original tests of their system in 1983 in Chicago, capitalized this name. Ericsson, developer of the çellular system in Scandinavia and now installing systems world-wide, uses the same acronym (MTSO) but expands the meaning a little: Mobile Telecommunications Switching Office. In the Aurora system, the developer, NovAtel, calls it a Master Mobile Switch and OKI calls it a Network Control Switch or, like Ericsson, a Mobile Telecommunications Office. Motorola's switch is called the Electronic Mobile Exchange (EMX). In trade journals and literature, the words "digital central office switch" and "cellular switching office" have also been used.

The most descriptive term seems to be "mobile telecommunications switching office" and will be used throughout this book as a generic term. It will not be used to refer to the Ericsson or AT&T equipment unless it is so stated.

There are three cellular system designs around the MTSO. The first type is the centralized design where the MTSO controls the entire system from one central site. A centralized system, from an OKI manual, is illustrated in Fig. 4-9. In this configuration the MTSO does the whole operation, the technical jobs as well as the administrative tasks, and is the interface with the local PSTN.

The second type is the decentralized system, in which there is a centrally located computer that handles the nonrealtime functions such as billing and traffic analysis. This sort of work requires a computer with somewhat more storage capacity. Technical jobs such as the actual communications and organizing hand-offs, for example, are handled by a number of smaller MTSOs at remote sites. They perform these functions independently of the main MTSO. This system is typified by the Aurora system (Fig. 4-10) by NovAtel.

Another configuration (Fig. 4-11) devised by the ITT/E.F. Johnson group under the name Celltrex, shows a division of the functions of

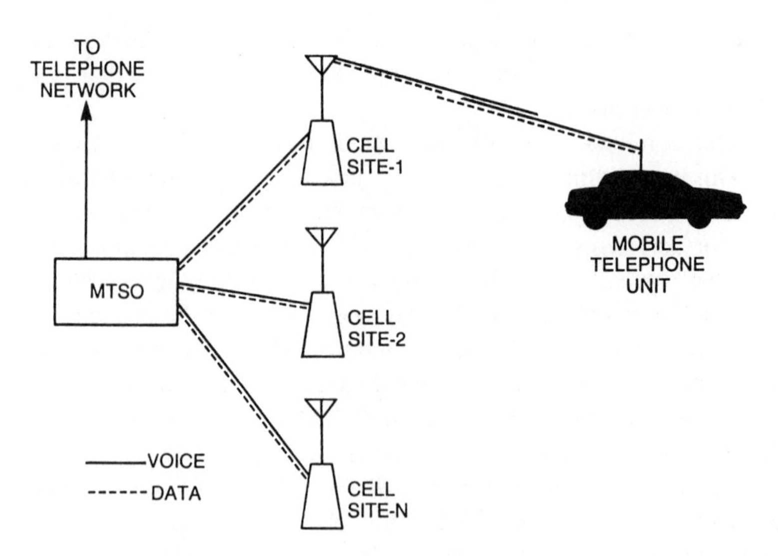


**Figure 4-9.** A typical centralized system.

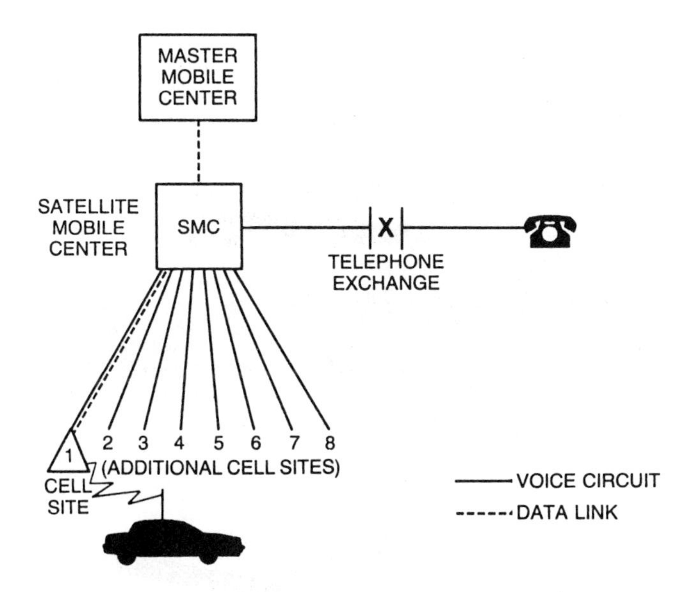


**Figure 4-10.** A decentralized system. (Courtesy of NovAtel Communications, Inc.)

the MTSO. The nontechnical tasks such as billing are done by the Network Control System (NCS), and the technical end of the business is shared by the NCS and the Remote Switch Group (RSG). The RSG also provides the interface with the local PTSN. It is claimed that with this method there is no need to route all of the calls back and forth through the MTSO or the NCS.

The third and last type is, as is to be expected, a combination of the first two types. An example is illustrated in Fig. 4-12, again a version of the Aurora system. This type is configured to fit a situation where a switch might be the central control for a smaller community but would also be connected as a remote site for a larger system nearby. An example could be a suburban town in Connecticut that would have a sufficient population to justify having its own system for local calls, but would generate enough business to need to connect into the New York area cellular system.

For a topic that has been stressed as being the "guts of the system," we seem to have dismissed it with a relatively short treatment, almost as though the MTSO is only a minor item in the system. Watching a switch/computer combination is, to use an often-quoted Irvin Cobb simile, ". . . like watching grass grow." The room will be

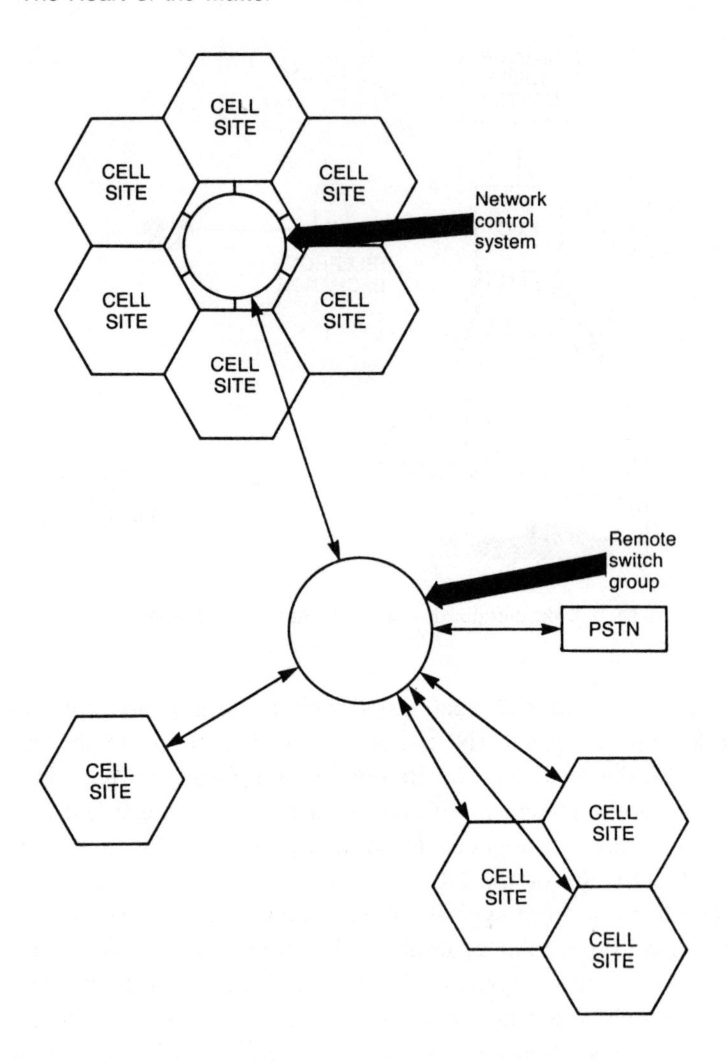


**Figure 4-11.** Another decentralized configuration. (Courtesy of E.F. Johnson Co.)

quiet and there will be nothing to indicate the millions of operations that are taking place except for a few small flashing lights.

The MTSO is like a lot of today's technology—it does what it is supposed to do quietly and efficiently and without very much human intervention.

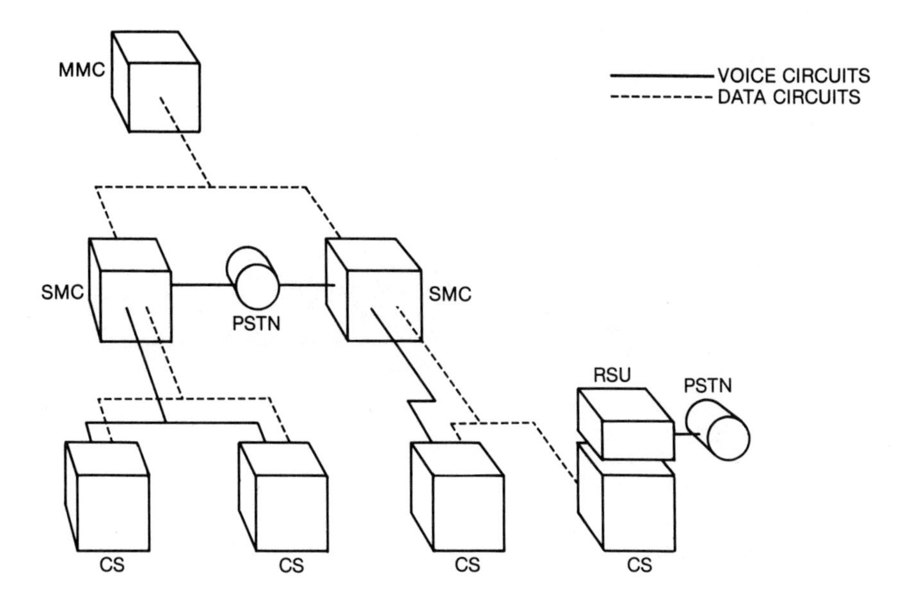


**Figure 4-12.** An example of the combination system. (Courtesy of NovAtel Communications, Inc.)

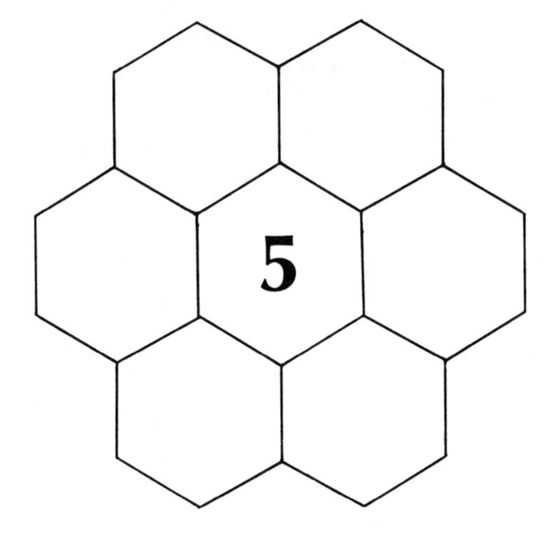


# The Cell Site

Although the obvious meaning of the term "cell site" is "the location of the cell," in the cellular industry the term has come to be used to include and describe the RF equipment located at the site. And so, when these words are used, they encompass the antenna, the tower or building upon which the antenna is mounted, the building or shelter housing the equipment, and the equipment itself.

The cell site antenna, distinguishable from other communication antennas by its configuration, is the subscriber's link to the rest of the system. At the base of the antenna is the equipment that is the interface between the mobile and portable units used by the customers and the local telephone system via the MTSO.

The equipment at the cell site does a number of tasks. These tasks are controlled by programs that are hard-wired into the equipment and by computers at the MTSO. As has been said, it provides the communications link with the subscribers and performs all of the tasks associated with this job such as setting up and terminating the calls, performing the hand-off routine, doing system testing under the command of the MTSO, and other miscellaneous jobs. In addition, the cell contains the redundant equipment that can be switched into the system in the event of failure of a component.

The physical location of the cell will be discussed in a little more detail later on in the chapter. You will see that the planning that must go into the process of setting up a system is very important. The FCC requires the hopeful system operator to go into considerable detail about the location of the cell and the expected operating characteristics of each cell. Since each cell can cost up to half a million dollars or more, the efficient design of the system is extremely important. When the operators need to enlarge the system to take care of additional subscribers and need to "densify" the system, they hope to be able to provide more capacity.

Figures 3-11 and 3-12 show an idealized example of the process of cell splitting. Figure 5-1 shows the same illustration with only one of the original seven cells enlarged or split to take care of additional subscribers in this particular area. This can be done by adding RF equipment to the original equipment in the cell for additional voice

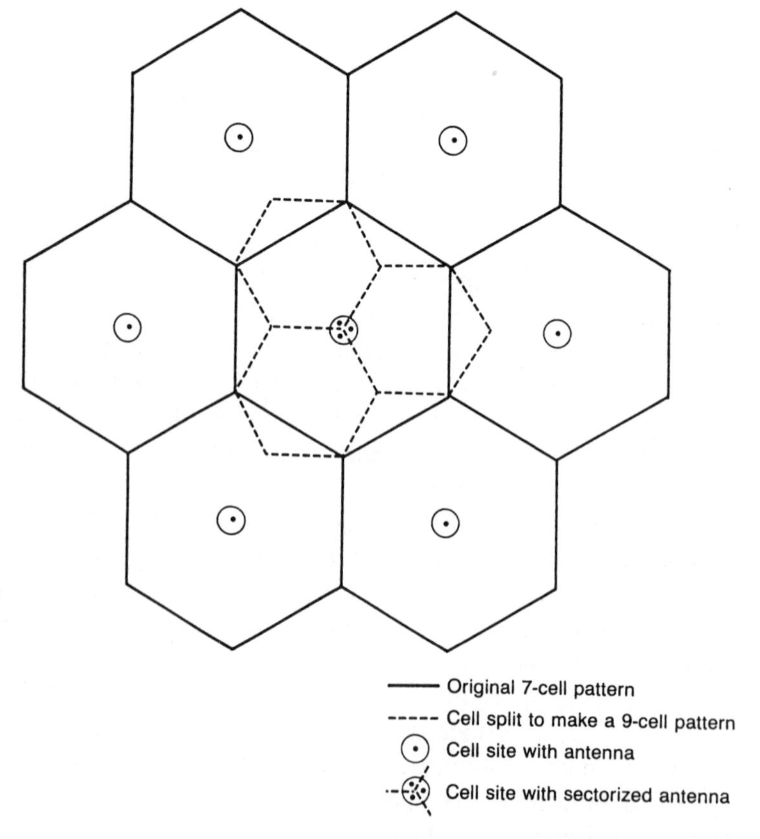


**Figure 5-1.** Idealized example of cell splitting to provide additional capacity.

channels and by "sectorizing" the antenna configuration to concentrate the RF radiation within 120° segments around the site. When this technique is used, the cell site is located at one corner of the cell rather than at the center, as has been shown so far.

The area that is to be covered by a cellular system is referred to officially as the Cellular Geographic Service Area (CGSA). The CGSA is located in a Standard Metropolitan Statistical Area (SMSA). The SMSA is a term used to define and outline areas that meet certain criteria established by the government agency that has the responsibility for developing this kind of criteria. From this comes data that is used in the production and analysis of population and demographic studies. The SMSA generally, but not always, follows county boundaries. It always contains an urban area together with the surrounding communities that have a close economic and social relationship with the central town or city. The data used to define these areas comes from computer analyses of census figures. There are about three hundred areas in the U.S. that have been defined at this time.

There are a number of factors that are considered when establishing an SMSA. Examples are population, family data, income, industry, etc. Boundaries of the SMSAs are not fixed forever because of the constant population shifts and changes that take place. Advertising and marketing people use this information routinely, and these areas have become known as *markets*.

The SMSAs, or markets, are ranked by size: New York, 1; Los Angeles, 2; Chicago, 3; Oklahoma City, 45; the Alton–Granite City market in southern Illinois, with a population of 20,500, 305. The FCC used these same standards to define the CGSAs, and in most cases the area and boundaries of the CGSA follows the SMSA. The markets were grouped in segments of thirty in accepting cellular licensing applications and, as of the publication date of this book, applications for licenses for the markets up to 180th are under review by the FCC.

Now that we have defined some of the terms with which we will be dealing, let's look at what is involved with locating the cells for these systems. Figure 5-2 shows one type of map that is used in initial engineering studies. This is a computer-generated schematic map, with the different area shadings indicating the varying density of population, household income, number of businesses and employees, and other factors that go to make up the SMSA. Overlaid on this data is a preliminary projection of the cells. In this example, five cells are considered necessary, and the coverage of each cell is shown.

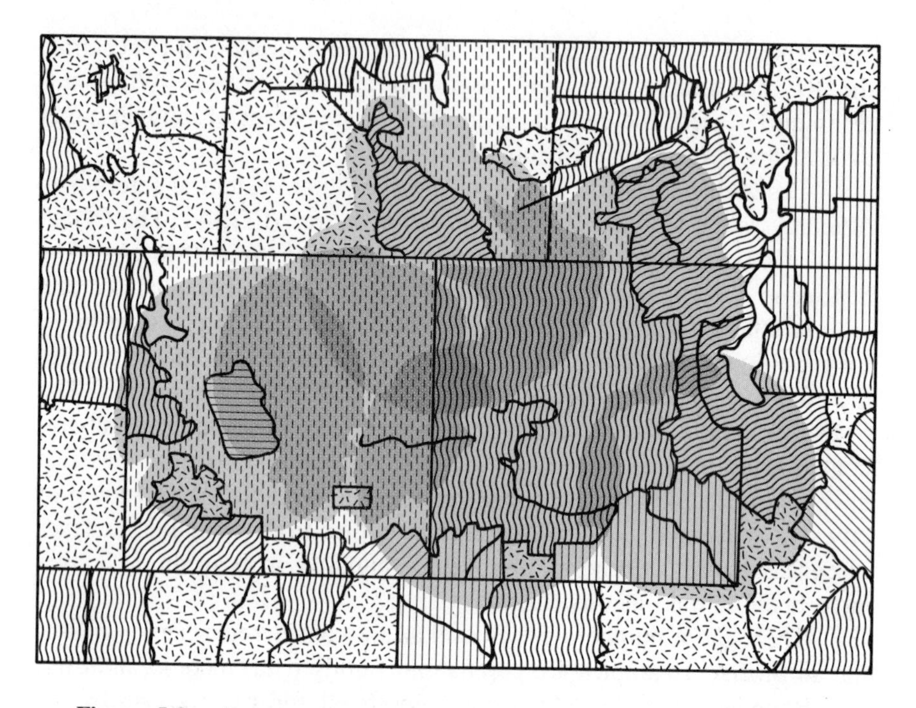

**Figure 5-2.** A computer-generated map for initial engineering studies. (Courtesy of Spectrum Planning, Inc.)

The illustration in Figure 5-3 shows a more detailed analysis of the projected RF coverage for another system. This propagation analysis is a prediction of the expected RF radiation coverage, again computer-generated, from the cell sites using data such as the geographic coordinates of the transmitters, the characteristics of the terrain, the height of the antennas, and the planned effective radiated power of the transmitter and the antenna. Figure 5-4 shows the cell coverage predicted for another SMSA. In the example, the predicted cell boundaries have been smoothed out by the computer.

Other factors have to be considered; one of these is the characteristic known as *in-building penetration*. Can a portable cellular telephone be used with satisfactory results inside the buildings? This is obviously very important in the urban, built-up areas that are expected to provide the major part of the market for the system. It is a big element in the system planning in the United Kingdom, discussed in Chapter 9, where portable instruments are expected to be a major part of the market in important areas such as London.

The rule of thumb in preliminary planning is to cover the heavily

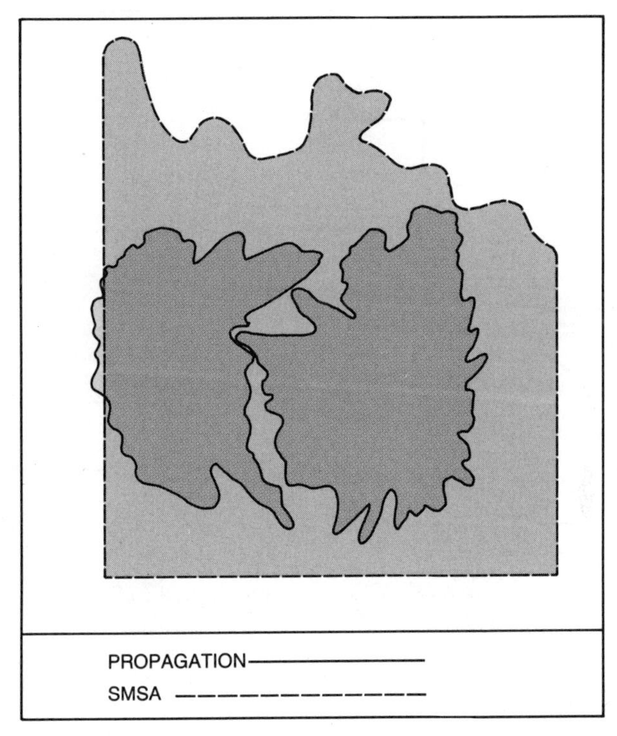


**Figure 5-3.** A typical propagation analysis. (Courtesy of Spectrum Planning, Inc.)

built-up areas having many multistory buildings with more than one cell. Another look at Figs. 5-2 and 5-4 will show you some overlapping areas of coverage for these planned systems.

When you consider factors that have already been mentioned and other factors such as the line-of-sight propagation path from each antenna and the multipath effects (effects that occur when some of the radiated power is reflected from an obstacle such as a building or a high hill), you can see how valuable the computer is in juggling so much data.

Small scale maps covering large areas, such as the United States Geological Survey (USGS) 1:250,000 series in which there are about 7800 square miles on a sheet, can be used for initial studies of the type that we have been talking about. The plots on Figs. 5-3 and 5-4 have been done on 1:250,000 maps. The final data, however, will probably come from the use of large scale maps such as the 1:24,000 sheets, which cover only about 60 square miles per sheet. At this scale, indi-

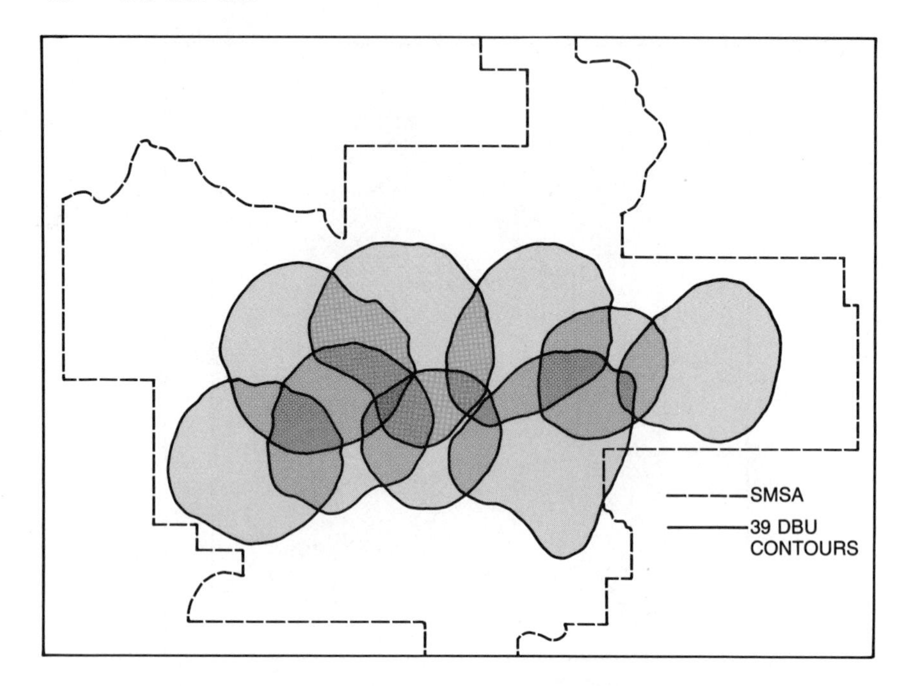


**Figure 5-4.** Computer-predicted cell coverage for one system. (Courtesy of Spectrum Planning, Inc.)

vidual buildings can be shown on the map and terrain elevations, or contours, are shown every 10 ft. On the smaller scale maps, the contour interval depends on the type of terrain but is frequently 50 ft.

Having done all this, the system must be looked at from the point of view of system balance. This means that you must ensure that the mobile instruments, with a power output of about 3 w, and the portables, with a power output of about 1 w, will be able to communicate efficiently with the cell site radios at the outer limits of the cell site's coverage. The propagation of mobile and portable radios is going to be the determining factor—there is no use designing the cell site for a large area of coverage if the mobile units can hear, but cannot answer, the calls directed to them.

One way to help this is by boosting the efficiency of the receive antennas at the cell site. A more detailed look at the antenna problem is given later in this chapter, but Figs. 5-5, 5-6, and 5-7 give some idea of one way this can be done and illustrate why you will see different types of antenna configurations at various cell sites.

In this example the transmitter propagation is omnidirectional meaning it radiates equally in all directions. This is the usual case.

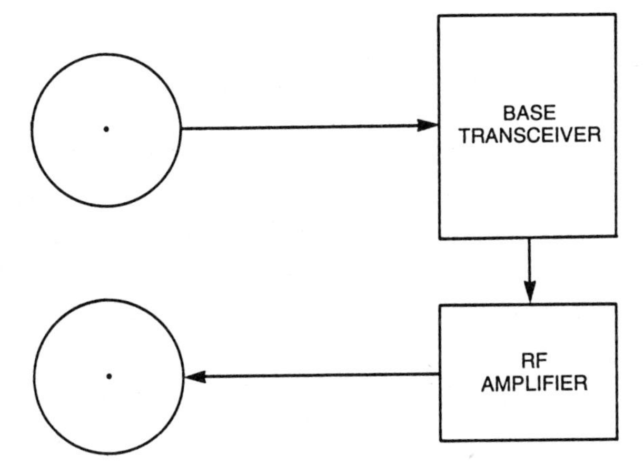


**Figure 5-5.** Omni-directional receive and transmit antennas. (Courtesy of E.F. Johnson Co.)

Figure 5-5 shows an example of a cell with a minimum number of available frequencies of channels and so the receiving antenna is omnireceive. When the number of channels has to be increased or there is a need to extend the coverage of the cell, four antennas can be arranged as shown in Fig. 5-6, and six antennas can be arranged as shown in Fig. 5-7.

In a cell site arrangement, as is shown in Fig. 5-7, where each receive antenna covers a relatively narrow area, at the cell site antenna there is likely to be a decrease in the signal strength received from a mobile that might move from an area covered by one antenna to

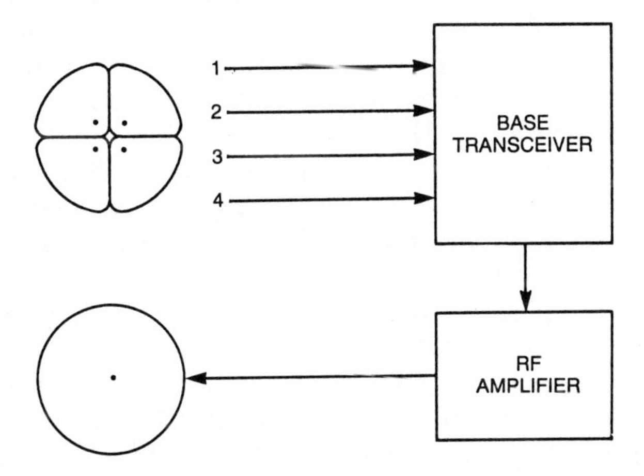


**Figure 5-6.** Four-sector receive antenna. (Courtesy of E.F. Johnson Co.)

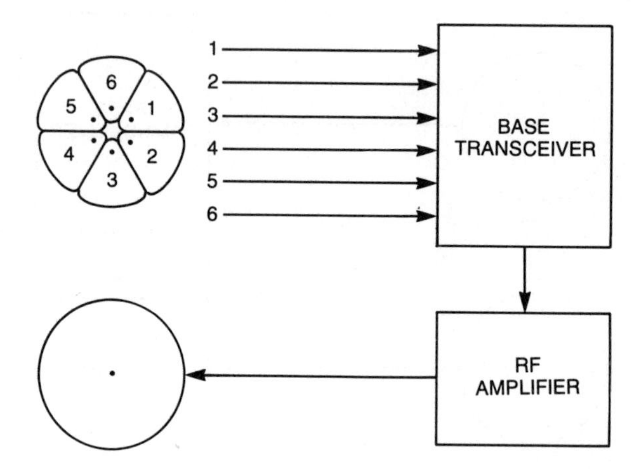


**Figure 5-7.** Six-sector receive antenna. (Courtesy of E.F. Johnson Co.)

another area—from area four to area six, for example. In this case, the transceiver that is constantly monitoring the signal strength and comparing it with the one received in the other antennas will switch the voice channel to antenna 6. This will result from a command on the part of the software in the logic unit at the cell site transceiver and does not require any action by the system operator who might be monitoring the system. The users at either end of the conversation will not be aware that this change has taken place.

We have discussed a few of the technical problems that have to be taken into consideration. In addition to all these problems, there is the problem of the physical location of the cell site—the antenna and radio equipment. Here there are factors that cannot be plugged into a formula and run through a computer. In many cases you are dealing with people's sensibilities, impacts on the environment, and already-existing local zoning regulations. These are the things that may force the system's design engineering staff to select something other than an ideal location for a cell site when only the engineering point of view is considered. It may require additional cell sites, at a considerable cost, to get the coverage that is wanted.

The main problem, of course, is the antenna support. No one wants a one-hundred-fifty-foot tower in the neighboring lot, although one may very well want a cellular radiotelephone in one's car. Probably the easiest solution to this problem is to locate the antenna on the top of an existing building or tower providing that there is room, the structure is high enough, there is no interference with other radios already on the location, and so on.

If a building can be used, then the radio equipment can be placed close to the antenna in a rooftop penthouse or in leased space on the top floor beneath the roof. With this type of installation, there will be no long cable runs with attendant power losses from the radios to the antenna. There will be few, if any, problems with zoning since there is already a structure on the site. This is particularly true in densely populated urban areas. Figure 5-8 illustrates this solution to an antenna installation. The suburban office building is located at the edge of a major shopping mall and heavily travelled commuter routes, but most passers-by do not notice the antennas.

The cell site, or base station as it is called in some systems, as we have said, contains the radio transceivers and the control and monitoring accessories needed to support the radio channels assigned to that cell. In addition, the equipment to maintain communication with the MTSO via leased landline or microwave relay is located here. A simplified block diagram of a typical cell site is shown in Fig. 5-9. This is the Ericsson CMS 8800 system.

The functions of the cell site equipment were described in some detail in Chapter 3. To review briefly, the cell site handles the radio communications with the mobile and portable subscribers who happen to be in that cell, monitors and supervises the quality of communications, and maintains a connection to the MTSO.

For each of the voice channels, as you can see in Fig. 5-9, there is

**Figure 5-8.** Cell site antennas on a fifteen-story building.

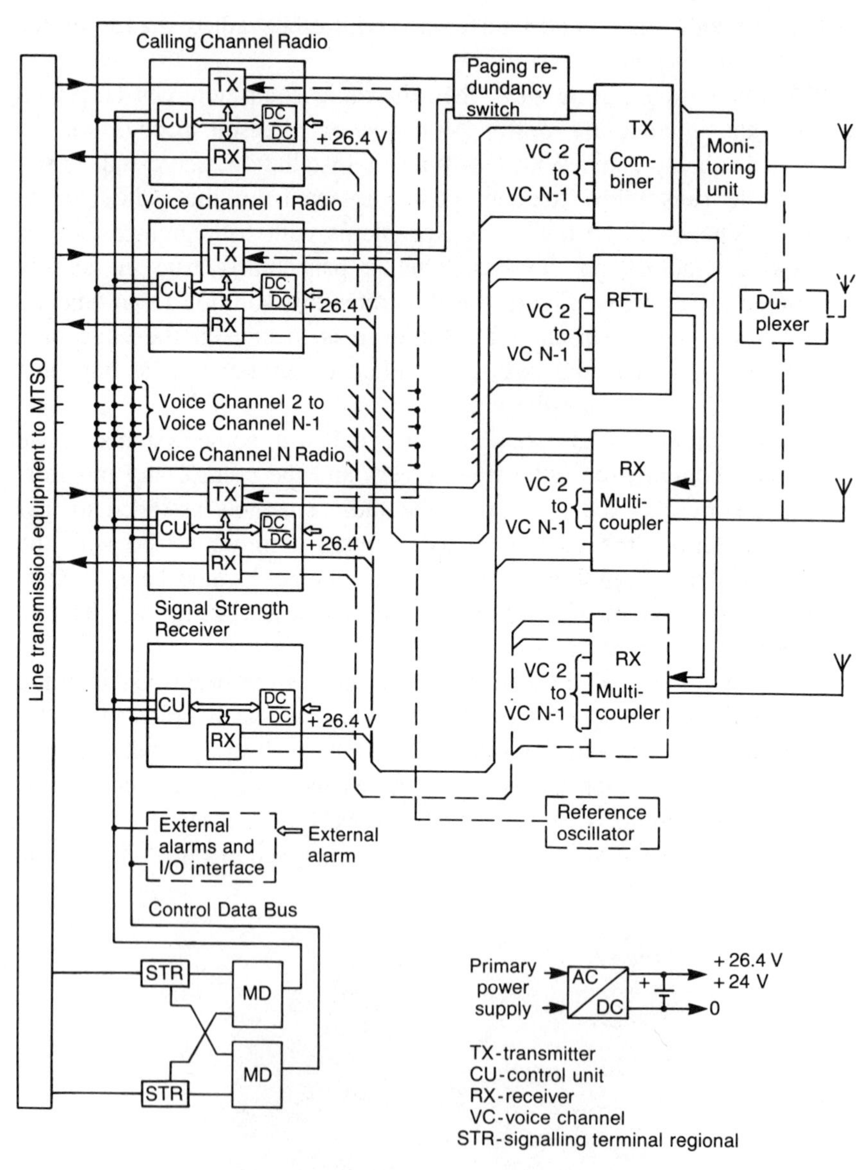


**Figure 5-9.** Cell site block diagram. (Courtesy of Ericsson Radio Systems.)

a module containing a transmitter (TX), a receiver (RX), a control unit (CU), and the power supply equipment (DC/DC) for these items. In addition, the cell site has a transmitter combiner and a receiver multi-coupler that are designed to allow the connecting of a number of transmitters or receivers to the same antenna without creating any interference to each other.

The signal strength receiver does only what the title says—it measures the strength of the radio signal on any of the channels on the command of the MTSO. This is done constantly to determine if the signal quality or strength of a call in progress has changed or deteriorated to the point where a handoff to another channel or cell site should be accomplished.

Let's look at this equipment in a little more detail, beginning with the CU. Figure 5-10 shows a block diagram of this component. This is essentially a dedicated microcomputer and, as the diagram in Fig. 5-9 shows, there is one for each radio channel and one for the signal strength receiver.

Like any computer, the CU must be told exactly what to do. Instructions about how to act on the data that is received come from the computer memory. The data that the CU acts upon comes from a variety of sources such as the subscribers, the MTSO and, of course, internal sources. This is the information that the control unit needs to accomplish its monitoring and supervisory tasks.

You can see now that it is obvious that the name describes the functions. The CU is responsible for all of the communications, both data and voice, with the MTSO, and it controls and monitors the dealings with the mobile and portable subscribers units.

The block diagram for the cell site transmitter is given in Fig. 5-11. The first block is the frequency generator that, acting on instructions from the CU, synthesizes the desired channel frequency in the transmit section of the cellular band, 870 to 890 MHz. This signal, at

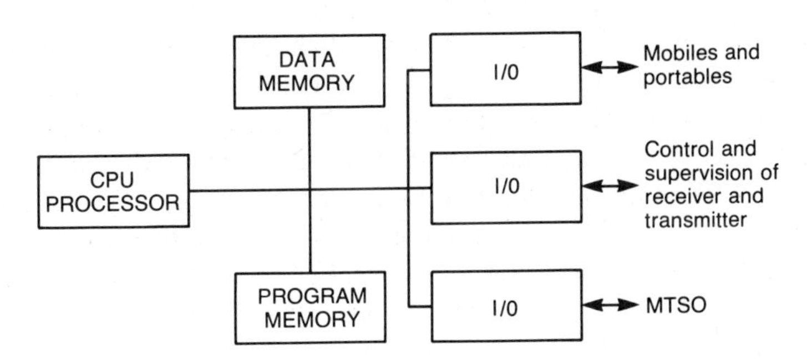


**Figure 5-10.** Central unit block diagram. (Courtesy of Ericsson Radio Systems.)

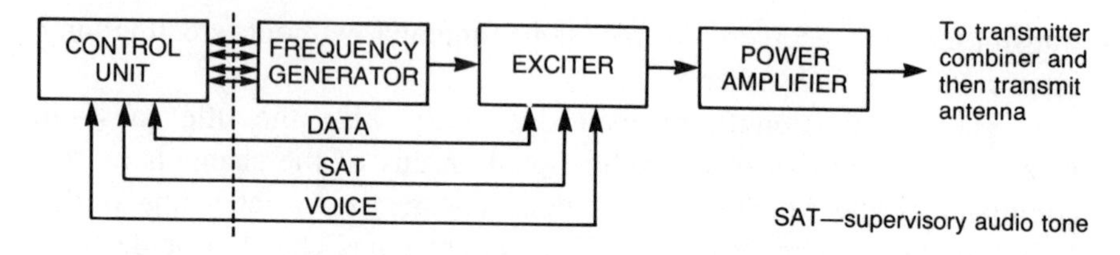


**Figure 5-11.** A transmitter block diagram. (Courtesy of Ericsson Radio Systems.)

the assigned frequency, then goes to the exciter, where the actual voice signal is impressed onto the frequency signal modulating it.

After some further processing and the initial amplification and filtering, the signal is sent to the power amplifier for the final step of amplification to the level set by the CU. You will recall that this power output is continuously monitored so that the signal strength is only amplified to the amount necessary to maintain satisfactory communication between the cell site and the subscriber.

Having gotten the transmitter signal this far, let's look at the final step needed to actually put it on the air before we look at the components of the receiver.

If there were only one communication channel and so only one transmitter in a cell site, there would be no need for the filter combiner or multiplexer. It is this component that allows more than one transmitter to be connected to a single antenna without any interference between them. Each combiner is tuned to one specific frequency—it will pass the RF energy only at this frequency and will block all other frequencies. This means, of course, that the energy from an adjacent transmitter that is assigned to a different frequency but connected to the same antenna cannot feed back into the power amplifier of another transmitter. Figure 5-12 is a schematic diagram of this type of circuit. The photograph in Fig. 5-13 shows the multiplexers installed in Motorola 8 channel cell site equipment. The multiplexers are the large tubular objects stacked four over four at the right of the photograph.

Now let's turn our attention to the receiver. Figure 5-14 is a block diagram of a typical unit. The incoming signal is routed to the correct receiver by the multicoupler. It is blocked from the other frequency channels by the action of these components—the action is essentially the same as in the transmitter combiners.

The signal goes into the high-frequency unit, where it gets the first stage of amplification and has the frequency shifted downward to that

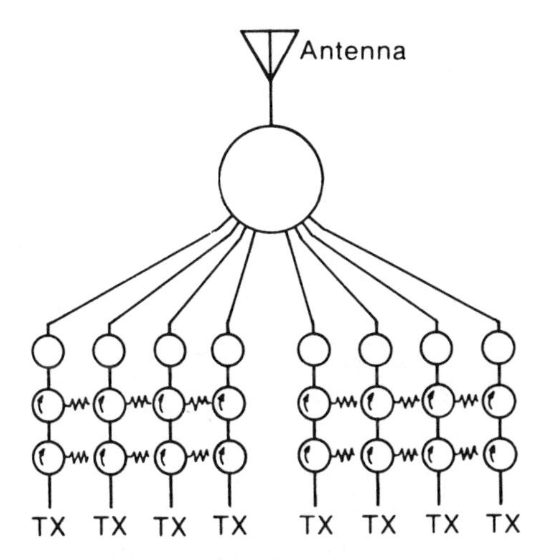


**Figure 5-12.** A typical filter-combiner schematic drawing. (Courtesy of Ericsson Radio Systems.)

of the frequency generator. Then the intermediate-frequency unit lowers the frequency of the signal still further and also feeds a signal strength report back to the control unit. The main portion of the signal then goes to the audio-frequency unit where it is converted to the audio-frequency range for transmission to the MTSO and then to the PSTN.

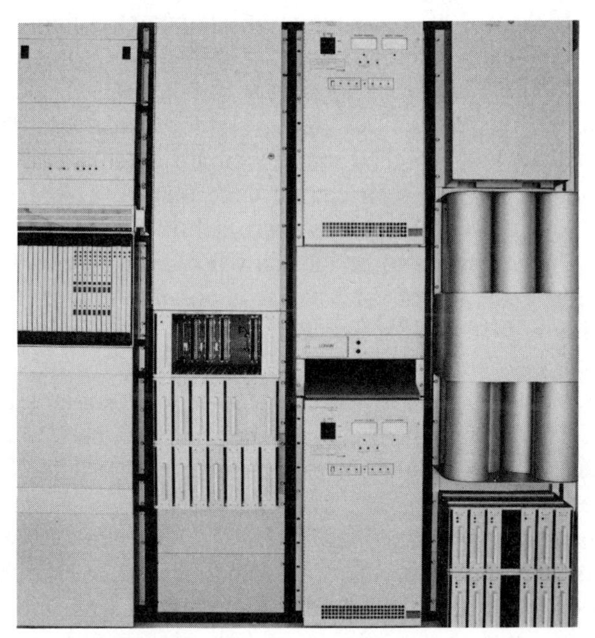

**Figure 5-13.** Motorola DYNA T-A-C cell site equipment. (Courtesy of Motorola, Inc.)

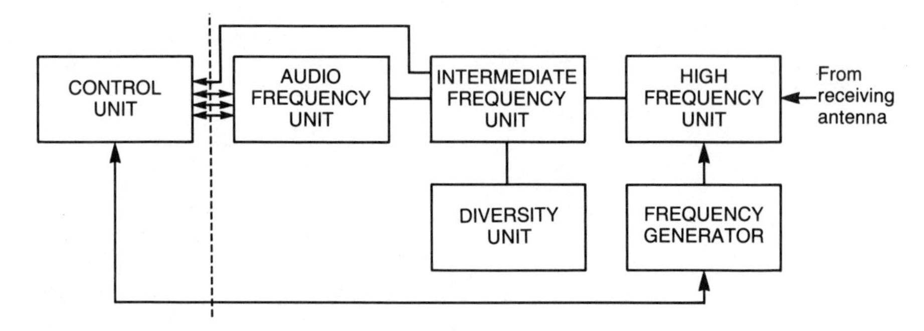


**Figure 5-14.** Receiver block diagram.

In the block diagram (Fig. 5-9) of the Ericsson system, two sources of power are shown for the cell site equipment. The first source is the main power supply connected to the commercial power lines that are run into the cell site area. In most cases this will be 208-V, 60-Hz AC power. This will be the power input to the transformer and rectifier system where it will emerge between 24 and 30 V DC. This will probably be "floated" across a bank of batteries that can be called upon to supply the necessary power in the event of a failure to the commercial supply. If there is a large amount of equipment at the site requiring a lot of power—many cellular channels and perhaps a microwave relay station for communication to the MTSO in lieu of a landline—there may be a diesel or gasoline powered generator available at the site that can be started up and switched into the circuit automatically by remote control from the MTSO, or it can be done manually. A diagram for a typical cell site power supply is shown in Fig. 5-15.

When the cell site is to be located where commercial power is not available nearby, the use of solar power is possible in many parts of the country. Figure 5-16 shows the use of solar power for a microwave station. Such a system will use the power generated by the solar cells to charge the batteries from where the power will come for the radio equipment. This arrangement can also be used as a backup to keep the batteries charged that would power a cellular system if the commercial power failed.

The power, now at a somewhat lower DC voltage instead of AC, is routed to the equipment where needed. Each one of the transceivers has what is known as a DC-to-DC converter, in which the incoming power is converted to the variety of voltages needed in present-day electronic circuits, +12 V, −12 V, and +5 V.

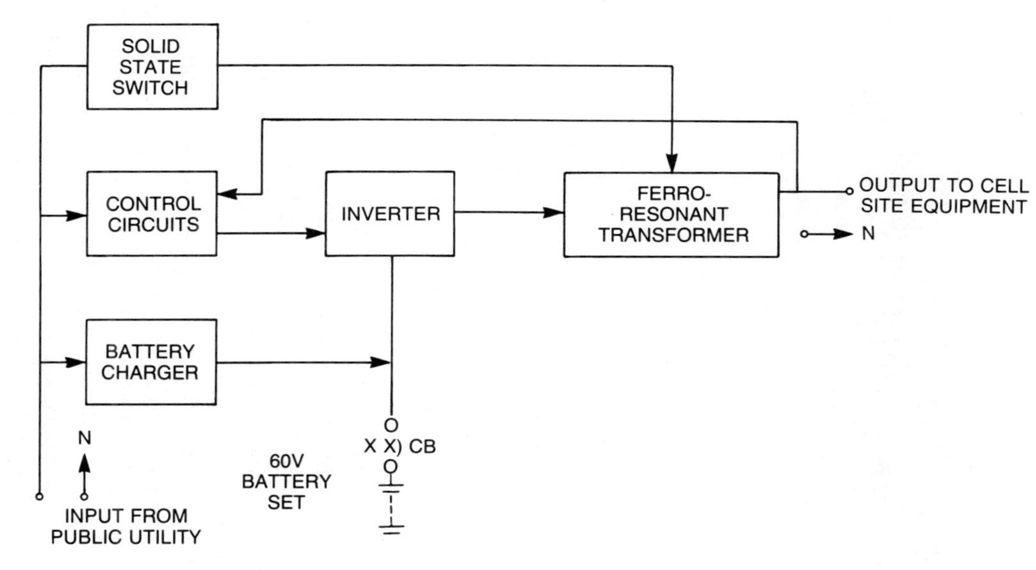


**Figure 5-15.** A simplified block diagram of a typical cell site power supply. (Courtesy of Exide Electronics, World-Wide Mfg. of UPS.)

**Figure 5-16.** Solar power for a cell site. (Courtesy of Photocomm, Inc.)

The operation of all the equipment, transceivers, power supplies, and so on, is monitored by circuitry built into each piece, and a failure or malfunction from anywhere will set off alarms at the MTSO. Not only will equipment failure trigger an alarm, but incoming power failure and physical damage to the equipment by vandalism will also cause bells to ring and alarm lights to light. In most of the cellular systems, the program in the main MTSO computer determines the cause for the alarm and displays it on a monitor screen or prints it out.

The maintenance procedures for the equipment follows along the same lines with a computer program built into the system and available on command to the technician on the job. Most cell sites are equipped with a transceiver similar to that used in a mobile unit by a subscriber. This is used as part of the testing program so that system faults can be duplicated at the cell site.

One last word on this topic: Although the Ericsson system has been used as an example—Fig. 5-17 shows the assembled Ericsson cell site radios with all of the components identified—the same block diagrams and descriptions, with minor modifications perhaps, could also apply to other manufacturers such as Motorola, Harris, and OKI, to name just a few. Figure 5-18 is a photograph of a partially removed transceiver and the remainder of the cell site radio by Harris. The transceiver shown in Fig. 5-19 is typical of the cell site radios used and is included to show the similarity with the mobile transceivers shown in the next chapter.

Let's go back a little now and again consider the physical location

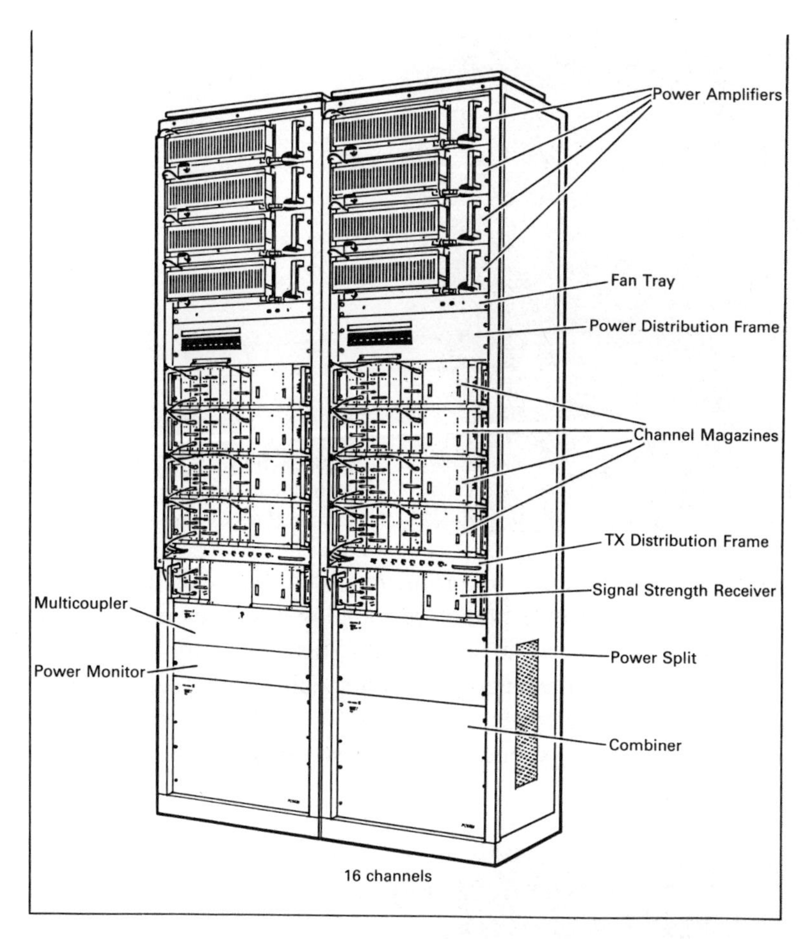


**Figure 5-17.** The assembled cell site equipment for the Ericsson CMS 8800. (Courtesy of Ericsson Radio Systems.)

of the cell site. It should be obvious that the cell site antenna must be raised some distance above the ground to radiate and receive the RF energy in a manner that covers the desired area for that particular cell. This can be done by placing the antenna on the roof of an existing building or tower, or on a tower built specifically for this purpose at the center of the cell. There are very few buildings over five stories high in today's urban areas that do not have some type of antenna on the roof, ranging from a simple omnidirectional vertical to a dish 10 m in diameter designed for satellite communication. As said earlier in the chapter, using an already in-place facility for an antenna is the preferred

**Figure 5-18.** The Harris Cellstar control station for cell site use. (Courtesy of Harris Corp.)

method that avoids a lot of problems, both financial and legal. But where a suitable structure cannot be found or just does not exist, then a support of some type must be erected. This is usually a tower.

Towers that are designed and built specifically to support antennas come in a variety of types and sizes. The self-supported monopole, shown in Fig. 5-20, is one of the most popular for use in cellular

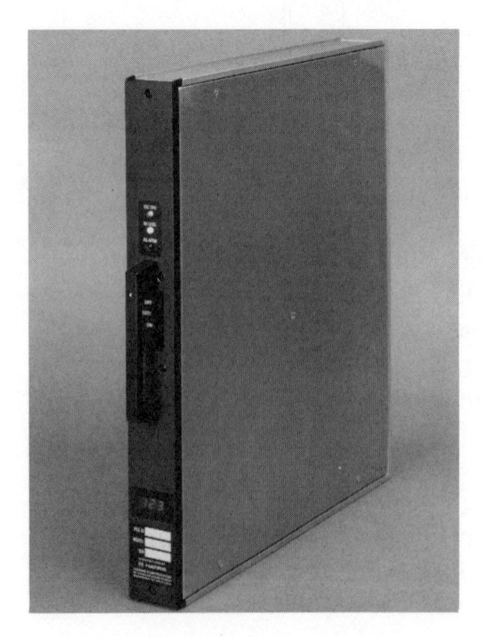

**Figure 5-19.** Transceiver used in Harris equipment. (Courtesy of Harris Corp.)

**Figure 5-20.** A cellular antenna support, a monopole.

systems, although it is probably the most expensive to buy and install. Generally about 150 feet in height, it is used in those locations where the antenna support is to be as unobtrusive as possible. It uses no guy wires and is almost always painted a light blue or green to blend with the horizon, sky, or surrounding trees. In the west in desert areas, a pale sandy color is used.

This type of tower is the easiest and quickest to erect—only one footing need be poured although it must be a big one—and it is the tidiest since the radio cables can be routed inside the supporting pole. Another advantage of the monopole is that this single support has the least effect on the radiated signal. There is also less of a problem with adventurous vandals since the rungs for climbing the tower can be easily placed and removed each time they are needed.

Figure 5-21 shows an example of a self-supported lattice tower in a rural area. This type can be almost any height and provides a somewhat steadier platform for the antennas than the monopole which will

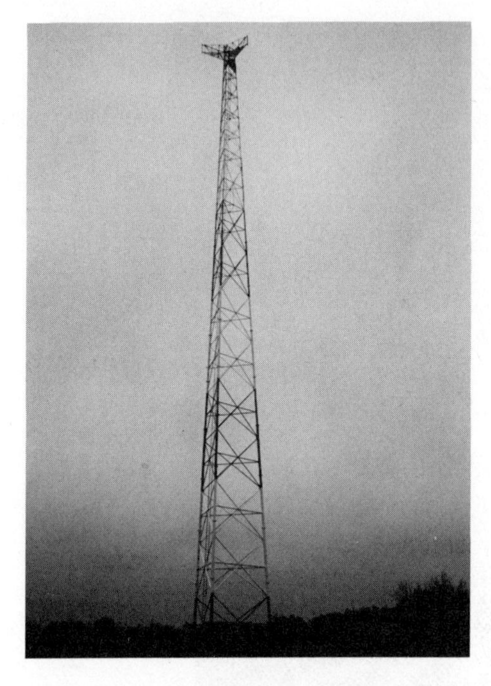

**Figure 5-21.** A self-supported antenna tower. (Courtesy of Microflect Co., Inc.)

sway a little in a high wind. Tower movement has little, if any, effect on the cellular signal. However, movement renders the monopole tower unsuitable for microwave signals because the extremely short microwaves act in the same manner as a beam of light and the antennas must be sited and aimed very accurately. A tower illustrating this particular point is shown in Fig. 5-22. Because of the number of microwave antennas on this tower receiving and transmitting signals in all directions, a stable support is needed. This is achieved by designing and building a tower with broad supporting sides and legs.

Let's go back now to the tower in Fig. 5-20. The vertical antenna at the very top of the tower is the transmit antenna, and just below this is the microwave dish antenna that provides the communication link with the MTSO. At a site like this, it is probably more economical to use microwave communication rather than to install a telephone line to this isolated area. Below the dish antenna are the sectored receive antennas.

The third type of tower is a guyed tower. Because of the supporting and bracing effect of the guy wires, the tower structure itself can be somewhat smaller than that needed in the self-supported towers. However, much more land is required since the guy wires are anchored at a

**Figure 5-22.** A microwave tower.

distance from the base that is nearly equal to the height of the tower. There are areas where a guyed antenna tower is essential because of the added support given to the structure by the guy wires; the Gulf Coast is one of these areas because of the frequent threat of hurricanes.

The engineering and technical problems of siting and erecting a cellular antenna support can sometimes be minor when compared with the difficulties in arranging for the lease or purchase of a suitable plot of ground and then getting the necessary zoning approval and building permits. In some cases the technical considerations must be subordinated to the aesthetic requirements imposed by the local zoning authorities. As an example, because the height of the towers was restricted by the zoning people to 100 feet rather than the usual and planned 150 feet, the cellular system in San Francisco needed two more cells to get the coverage of the entire area originally planned.

Let's look now at the radiating and receiving elements, the antennas, that are going to be installed on the tower. The exterior view of a typical antenna is that of a polyvinylchloride, (PVC) or fiberglass pipe ten or fifteen feet in length and two to three inches in diameter. The cable feeding the RF energy to and from the antenna elements and the cell site radios goes into the bottom of the tube. The synthetic housing shields the antenna from the effects of the weather, but it does not affect the radiation or reception of the antenna.

The actual antenna, which is the component that does the radiating, is usually a center-fed collinear type with two or more elements stacked one above the other to provide some gain or reinforcement of the signal. An idealized example is shown in Fig. 5-23.

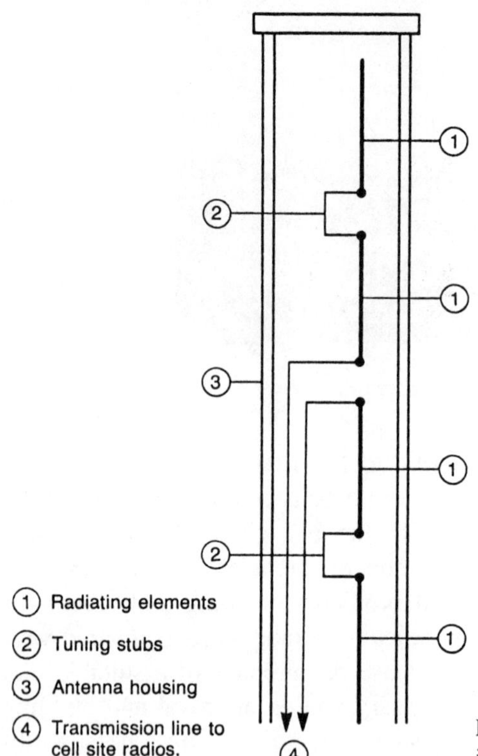


**Figure 5-23.** Schematic of an idealized collinear antenna.

The energy radiated from a cellular antenna in the transmission mode is vertically polarized for a number of reasons. One reason is that mobile antennas are mounted vertically, and greater efficiency is achieved when both antennas point in the same direction.

The vertical radiation pattern of a typical omnidirectional cell site antenna is shown in Fig. 5-24. These radiation patterns show receiving as well as transmitting characteristics. The horizontal pattern for an omnidirectional antenna will be essentially a circle around the antenna elements. When a reflector is added, as in the case of a sectorized receiving antenna, the pattern is changed, as illustrated in Fig. 5-25. Notice the difference in the reception characteristics of the antenna caused by the installation of the reflector.

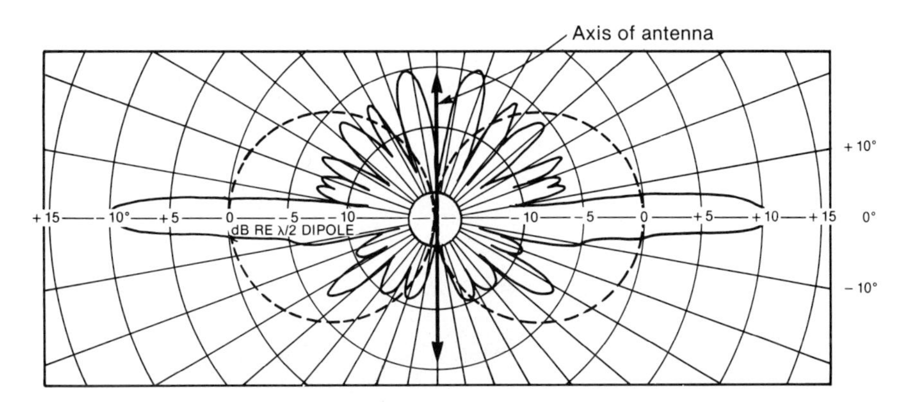


**Figure 5-24.** Vertical radiation pattern of an omni-directional antenna. (Courtesy of Sinclair Radio Laboratories, Inc.)

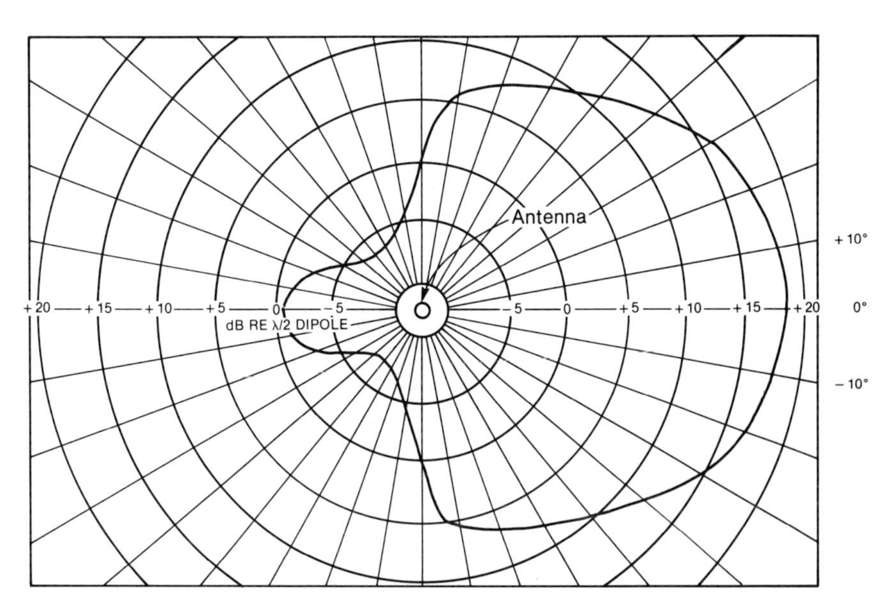


**Figure 5-25.** Horizontal radiation pattern of a sectorized antenna. (Courtesy of Sinclair Radio Laboratories, Inc.)

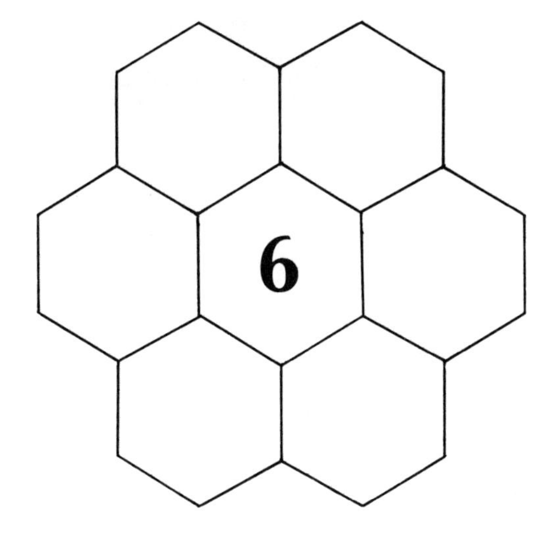


# The Mobile Station

Now we come to what is probably the only part of the cellular mobile radiotelephone system with which the subscriber comes in contact—what one manufacturer refers to as the '"vehicle-mounted subscriber radio unit". Here, however, the term *mobile station* includes the so-called portable and briefcase units as well a those mounted in vehicles. In most cases it is just a question of packaging—the electronics and operational procedures remain the same in all the units.

The two most visible components of the cellular system are the antenna on the roof of the vehicle or on the lid of the trunk, and the handset and cradle in the driver's compartment. In the trunk or under the seat of the car, however, is the transceiver/logic unit, that is, the black box that controls and monitors the radio communication with the cell site.

All the components that make up the mobile station, no matter who manufactures them, operate in essentially the same manner. They would have to do this since they are operating in a rather tightly controlled and regulated environment. The specifications for this equipment and, in fact, for the whole cellular system, insures that you can have any manufacturer's unit in your car and it will communicate satisfactorily with the equipment in the cell site no matter who manufactured its equipment. Too, like so much of today's electronic equipment, there is no guarantee that the name or logo on the nameplate of an item is that of the company that manufactured it.

Let's look now at each of the components in the vehicle and see what part they play in the system and how they do it.

The handset or control unit comes in a variety of configurations to suit any requirement and Figs. 6-1 and 6-2 show two examples that are typical of what is available. These are the units before they are mounted

**Figure 6-1.** One of Motorola's line of handsets. (Courtesy of Motorola, Inc.)

in a vehicle or in a briefcase, and the mount for the instrument cradle is not shown. These mounts can vary a good deal depending upon the desired arrangement in the driver's compartment.

There are several types of handsets manufactured. In one model, the keypad, controls, status indicators and display are located on the cradle. Another model has the controls and indicators on the back of the handset, and is somewhat larger but still not too bulky to handle easily. The Motorola model shown in Fig. 6-1 has the less frequently used controls in the cradle and the more frequently used controls, like the keypad, on the handset.

One of the better charts describing the purpose for the controls, buttons, switches,and indicator lights is given in the OKI instruction manual that comes with the instrument. Figure 6-3 shows a unit in

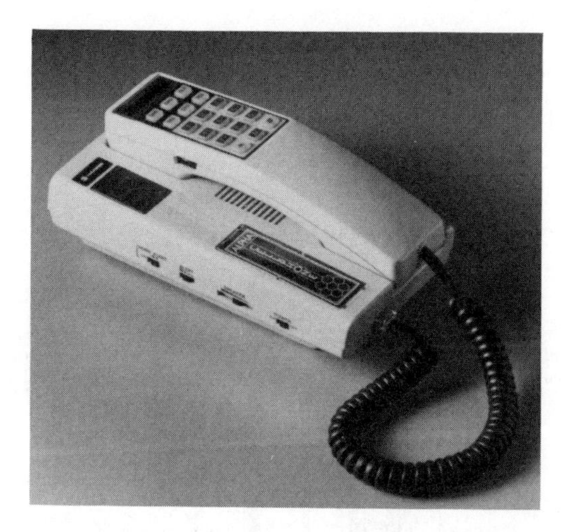

**Figure 6-2.** A Harris handset. (Courtesy of Harris Corp.)

which all the controls and indicators are in the cradle or base, and the handset is mounted at the side. Another style with the handset mounted on the front of the base is shown in Fig. 6-4.

What is involved in making a telephone call with these instruments and how is it different from the telephone in your home or office? Probably the first difference is that you do not have to go "off hook" as the telephone people say, to initiate a call. You can leave the handset in the cradle if it is more convenient for you. The numbers or digits that you enter into the keypad are stored in a memory IC or chip in the handset; they are not transmitted yet. As a check, the numbers are displayed in the area above the keypad so that you have a chance to review what you have punched into the unit before it is transmitted.

After you review the number on the display and are sure that it is correct, you push the SND (send) button and the wanted number is transmitted via the transceiver and antenna to the cell site. From here, of course, the signal goes to the MTSO and to the local telephone lines.It is then treated like any other telephone call and routed to the wanted subscriber. You will recall from the description of the system in Chapter 3 that the mobile unit is always in contact with a cell site and that a constant stream of data concerning the status of the unit and quality of the communication path is being sent back and forth.

Now the telephone is ringing at the number you dialed. You can hear this signal through the speaker in the cradle or the handset, and

**THE DIRECT LINE™**
**Controls and Indicators**
**Dial In Base Model**

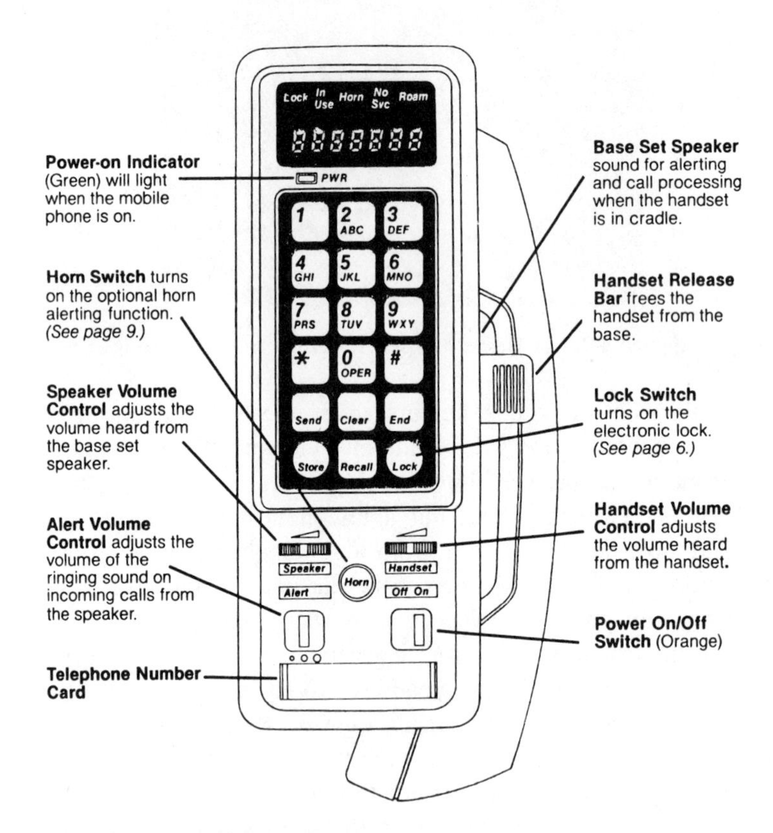


**Figure 6-3.** An OKI chart explaining the functions of each of the controls. (Courtesy of OKI Telecom, Cellular Telephone Division.)

you have not, so far, done anything other than dial in the number of the party you wished to reach and pressed the SND button. You have not picked up the handset.

If the party you want answers the call, you have two choices to continue. You can either pick up the handset and use it like a conventional telephone or, in some installations, you can push an appropriate button and carry on the conversation through a remote microphone and

# *THE DIRECT LINE*™
## Controls and Indicators
### Dial In Handset Model

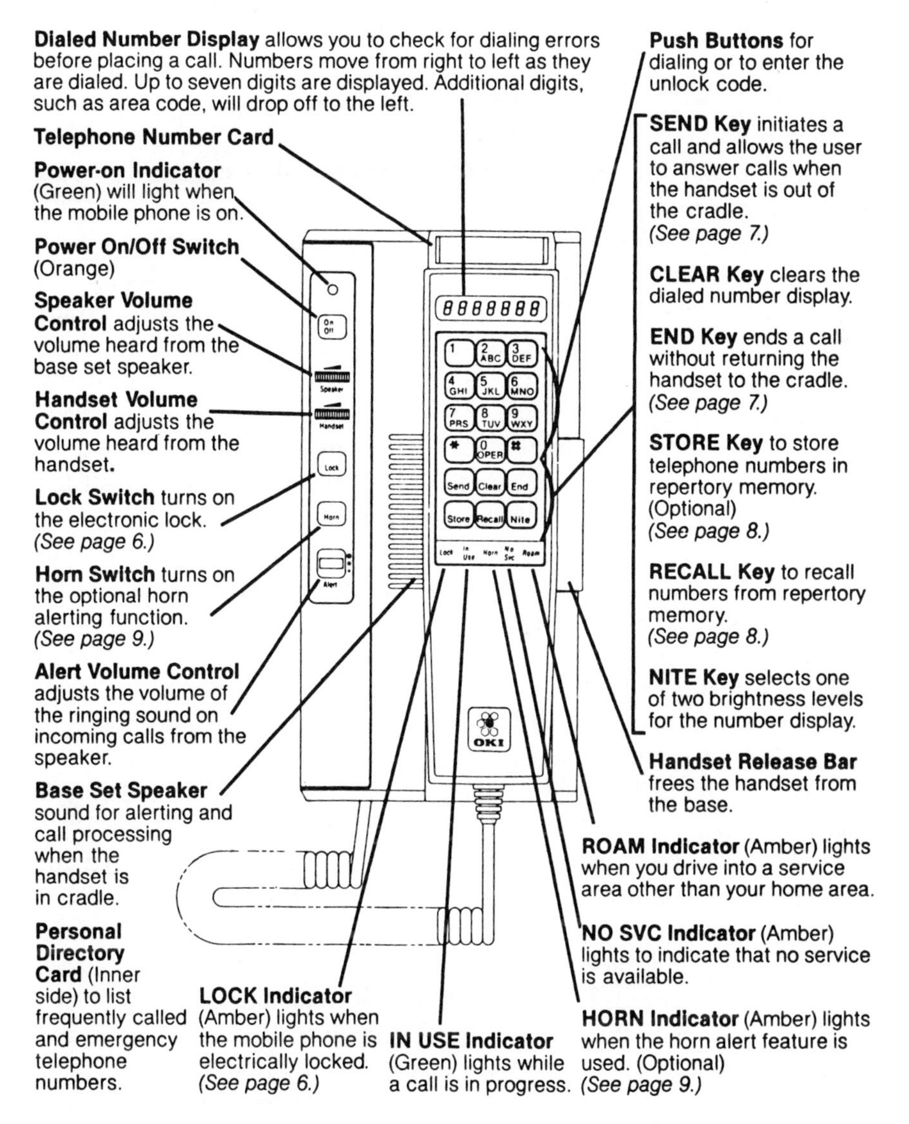

Dialed Number Display allows you to check for dialing errors before placing a call. Numbers move from right to left as they are dialed. Up to seven digits are displayed. Additional digits, such as area code, will drop off to the left.
Telephone Number Card
Power-on Indicator (Green) will light when the mobile phone is on.
Power On/Off Switch (Orange)
Speaker Volume Control adjusts the volume heard from the base set speaker.
Handset Volume Control adjusts the volume heard from the handset.
Lock Switch turns on the electronic lock. (See page 6.)
Horn Switch turns on the optional horn alerting function. (See page 9.)
Alert Volume Control adjusts the volume of the ringing sound on incoming calls from the speaker.
Base Set Speaker sound for alerting and call processing when the handset is in cradle.
Personal Directory Card (Inner side) to list frequently called and emergency telephone numbers.
LOCK Indicator (Amber) lights when the mobile phone is electrically locked. (See page 6.)
IN USE Indicator (Green) lights while a call is in progress.
Push Buttons for dialing or to enter the unlock code.
SEND Key initiates a call and allows the user to answer calls when the handset is out of the cradle. (See page 7.)
CLEAR Key clears the dialed number display.
END Key ends a call without returning the handset to the cradle. (See page 7.)
STORE Key to store telephone numbers in repertory memory. (Optional) (See page 8.)
RECALL Key to recall numbers from repertory memory. (See page 8.)
NITE Key selects one of two brightness levels for the number display.
Handset Release Bar frees the handset from the base.
ROAM Indicator (Amber) lights when you drive into a service area other than your home area.
NO SVC Indicator (Amber) lights to indicate that no service is available.
HORN Indicator (Amber) lights when the horn alert feature is used. (Optional) (See page 9.)
OKI


**Figure 6-4.** Another OKI model. (Courtesy of OKI Telecom, Cellular Telephone Division.)

speaker generally mounted on the sun visor. There is no need to use the handset at all.

When the call is completed, you replace the handset, if necessary, and push the END (end) button. If you have been using the remote microphone, you need only push the END button.

If for some reason you cannot complete the call, you end the call using the procedure just described. When you want to try the number again, all you need to do is to press the RCL (recall) and # button. The last number you dialed will be displayed and ready for you to press the SND button and try again. There is no need to redial since the number is still in the memory.

Mounting the handset and cradle conveniently in the driver's compartment is no great problem. Where the space between the front seats and the firewall contain the transmission controls or a radio or tape player, a bracket can be suspended under the dashboard and the cradle and handset attached to the bracket. Some installations can be made by putting the cradle in the space between the seats on the drive shaft tunnel or, in the case of front-wheel-drive cars, it can be mounted on the floor. There are some dashboard arrangements that will permit the vertical mounting of the cradle on or under the dashboard.

In any case, you may be sure that the dealer doing the installation will be able to fit the equipment into a convenient place where it can be easily reached and handled by the driver.

There are generally two sets of flexible rubber-covered or plastic-covered cables coming from the cradle, one to connect with the transceiver/logic unit and the other to connect with the power source. The cables are small enough so that they can be easily run under the floor mats or up under the dashboard where they are hidden and do not interfere with the operation of the car.

The preferred location from which to take the electrical power for the unit is an unused position on the vehicle fuse panel. This panel is generally located under the dashboard. The power to the panel comes directly from the vehicle battery and, because of the relatively short length of cable needed, the voltage drop is at a minimum. The wire that carries the power to the transceiver runs along with the cable carrying the communications from the handset back to the transceiver.

If the fuse panel is not convenient, or if there is not an unused position on the panel, the power can be taken from the nearest source such as a cigarette lighter. In this case, an in-line fuse can be installed in the line for safety reasons.

The series of photographs in Figs. 6-5, 6-6, and 6-7 illustrate one type of cellular radiotelephone installation. Figure 6-5 shows the driver's compartment prior to the start of any work. It can be seen that

**Figure 6-5.** Car interior prior to cellular telephone installation.

there is a mobile radio already installed in the car. In Fig. 6-6, the cables have been threaded through from the fuse panel and the transceiver. The power cable is the one on the left and the communication

**Figure 6-6.** Cables from fuse panel and transceiver unit.

cable is the one on the right with the socket installed. The socket was installed at the factory and nothing needs to be done by the technician doing the installation.

The holes for the base plate of the mount are being drilled in Fig. 6-7. Sheet metal screws are used to fasten the base plate to the floor

**Figure 6-7.** Installing the mounting pillar for the cradle and handset.

boards. Figure 6-8 shows the installation practically completed with the cables yet to attach. The handle at the right side of the mount controls the position of the cradle and is tightened after the user has adjusted the unit to satisfaction. An installation such as this takes about 2 hours for an experienced technician and the time could be reduced considerably by the use of an additional technician.

**Figure 6-8.** Cradle and handset installed; connecting cable not plugged in.

The transceiver/logic unit (the black box that is hidden away) contains the electronic circuitry that transmits and receives data and voice information to and from the cell site. This unit performs two functions, as the name states and is really two units in one. The transceiver unit does the actual communicating back and forth with the cell site. It receives the voice and data signal associated with an incoming call and the constant monitoring, and it also transmits the data information and the voice of the subscriber. The other unit, the logic, contains all of the controls for the transceiver, memory, and microprocessor that runs the whole operation.

Both of these units are packaged in the same case, generally made from cast aluminum. Figure 6-9 shows the transceiver/logic unit mounted on the floor of the trunk of the car. The large cable coming from the top of the photograph comes from the control unit in the front of the car—it is the same installation that was described in the previous section. The smaller cable, just visible at the upper left corner of the unit, is from the antenna mounted on the lid of the trunk, and has been routed underneath the padding in the trunk.

The cooling fins cast into the side of the case are designed to carry away the heat that is generated by the components in the case. The unit is mounted so that the fins are exposed to as much air as possible. Figure 6-10 shows the mounting plate for this unit and the routing of the cables before the unit is mounted.

**Figure 6-9.** Transceiver with cables attached and mounted on mounting plate.

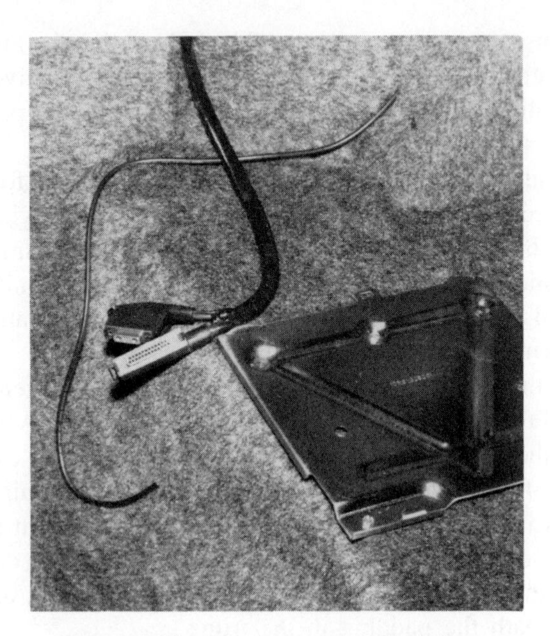

**Figure 6-10.** Transceiver mounting plate.

This same unit with the cover off on the maintenance bench is shown in Fig. 6-11. One of the circuit boards on which the components are mounted is easily seen. Across the front of the unit, from left to right, is the antenna connection, the lock with key that secures the unit to the mounting plate to help prevent theft, the fold-down handle for transporting the unit when necessary, and the two input sockets. The upper socket is for the control and voice and data lines from the handset/control unit. The lower socket is for the power from the source described earlier. There are no exterior controls since all the necessary adjustments for operation are made at installation when it is necessary to open the unit to install the individual calling and ID numbers.

The sizes of these units will vary, of course. One manufacturer has a unit $7\frac{1}{2} \times 9 \times 2\frac{1}{4}$ in. that weighs $5\frac{1}{2}$ lb, while another unit manufacturer has a unit approximately $10\frac{1}{2} \times 11\frac{1}{2} \times 2\frac{1}{2}$ in. that weighs $11\frac{1}{2}$ lb.

A simplified block diagram of a typical unit is shown in Fig. 6-12. This is an Nippon Electric Company model and shows how this manufacturer has divided up the functions performed by this piece of equipment. Each of the units in the case is contained on a printed circuit board that can easily be pulled out and replaced for servicing if

**Figure 6-11.** Interior view of transceiver unit.

this ever becomes necessary. Let's look at the functions of each one of these modules in a little more detail.

The users' instructions to the transceiver in the form of voice and data signals generated by controls and buttons on the handset come into the transceiver through a multi-pin plug. This plug terminates the cable running from the handset in the driver's compartment. The socket in the transceiver is wired so that it can be used as a connection for a diagnostic test set for trouble-shooting, maintenance, and adjustment purposes.

In the diagram for this model instrument, the power comes directly from the vehicle battery and is distributed by the DC power control unit to the individual circuit boards. This power control unit (Fig. 6-12) contains the remotely controlled on-off switch and it has the means to protect all the circuits from over-voltage (voltage surges and other electrical malfunctions that can occur). The power control unit in this NEC model also does audio processing, that is, preparing the speech signals and putting them into suitable form for transmission or reception.

The logic unit is designed to perform four functions. The first

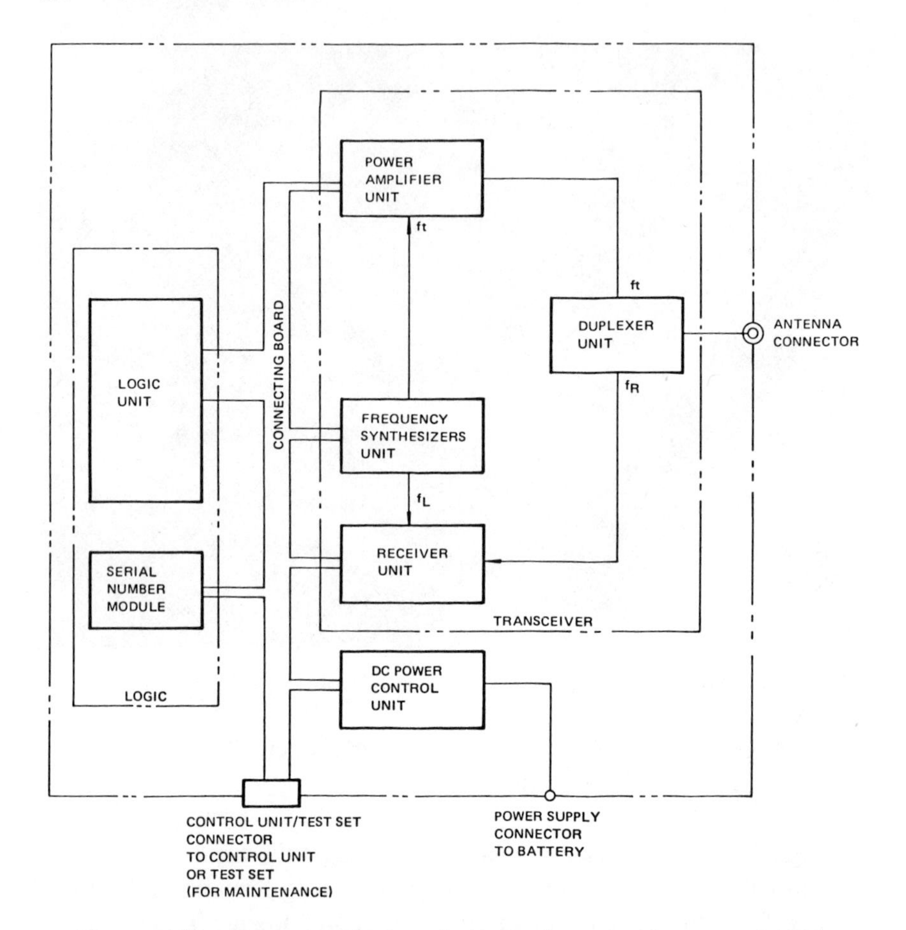


**Figure 6-12.** Transceiver/Logic unit block diagram. (Courtesy of NEC America, Inc., Mobile Radio Division, Hawthorne, CA.)

three functions are: Data reception and generation, control of the entire unit, and generation of tones such as the DTMF and the ringing signal you hear when someone wants to speak to you. The fourth function is the control of all the other functions, a sort of housekeeping task that is organized and supervised by the microprocessor.

The serial number module is also part of the logic unit. The module containing the Programmable Read Only Memory (PROM), the integrated circuit that has your individual telephone number and ID number *burned in* is also part of the logic unit. The telephone and ID numbers are put into the unit at installation with a reader/programmer operated by the installer. Unlike some types of ICs in which the information can be entered, stored, read at will and then erased and new information entered when necessary, the data on this PROM cannot be changed. As a consequence, if you ever want to change the telephone number assigned to your unit, the PROM must be removed from the transceiver and a new one processed and plugged in. This is a five minute task.

In Fig. 6-13, the PROM is the small grey object just under the Caution instructions. The complete reader/programmer is shown in Fig. 6-14. Data is introduced into the system by means of a keyboard and is displayed in the panel just above the keys. After the data is read and verified on the display by the technician, it is entered into the PROM by a simple flick of a switch.

The receiver portion of the unit is designed to receive any of the assigned 666 receive channels in the 870 to 890 MHz portion of the band. The requirements for this receiver are not as rigid as those for a high fidelity FM tuner that you might use in your living room. This cellular receiver is only required to reproduce the frequencies associated with the human voice that, as you will recall, are far lower than those associated with music. Also, the receiver is never very far from the transmitter, since the average cell is two to three miles in diameter. The

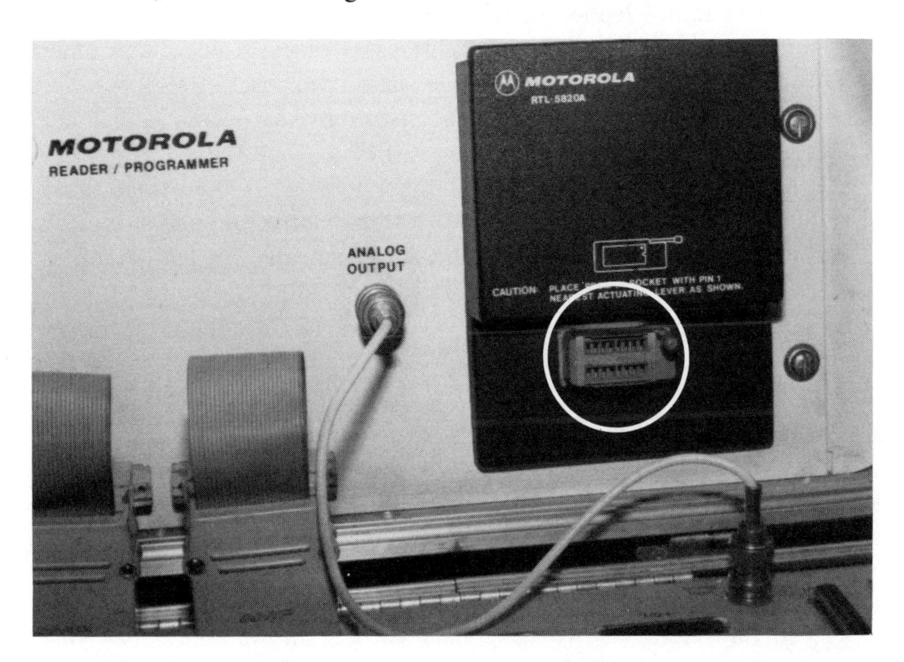


**Figure 6-13.** Motorola reader/programmer showing IC in position for "burning in."

**Figure 6-14.** Technician entering data in reader/programmer.

conditions under which the receiver has to operate, though, are somewhat more demanding than a home stereo set. The shock and vibration that occur in radiotelephone situations can be countered by mounting techniques learned though a number of years of experience with other mobile radiotelephone equipment. The same holds true of temperatures encountered and so the NEC unit specifications, for example, call for operation within a range of −30°C (−20°F) to +60°C (+140°F). For operations in climates where the temperatures will be below those specified, arrangements must be made to remove the unit and store it in a warmer place if the vehicle is going to be unheated for any length of time.

The frequency synthesizer, at the direction of the logic unit, generates the signal that is modulated and subsequently transmitted at the correct frequency/channel. This unit also contains the frequency modulator and the oscillator.

The power amplifier does just what its name says. It takes the completely processed signal from the synthesizer unit and amplifies it as much as necessary to reach the designated cell site. The signal strength, of course, is controlled by the logic unit that has, in turn, received instruction from the cell site.

The duplexers, as we have said earlier, allows more than one transmitter to be connected to the same antenna. This is the last unit in the chain before the antenna actually radiates the signal.

The antenna is the most visible component of a cellular mobile radiotelephone. It is instantly recognizable and is distinguished from all other types of antennas used with other mobile radiotelephone installations by the multi-turn curl at the base. These turns make up the air-wound phasing coil that permits the antenna to operate efficiently on any channel or frequency throughout the band. Unlike other mobile radiotelephones that receive and transmit on one or two channels that are reasonably close together, the cellular system must cover a relatively wide band for receiving (825 to 845 MHz) and another band (45 MHz wide) for transmitting.

The usual problems associated with antenna design to achieve maximum efficiency are compounded in mobile radiotelephone installations by the operating location. However, the engineer's task is made somewhat easier by the fact that the antenna is physically small. The antenna can be mounted in any number of locations on the vehicle to give an effective radiation patter. Unlike an eight foot whip antenna mounted on the bumper, the cellular radiotelephone antenna is not subject to fluctuations in the radiation pattern caused by flexing due to the vehicle's motion.

Antenna design is a much-studied topic in the field of communication and the literature is very extensive. Because of the way RF energy flows and is radiated, antenna dimensions are generally referred to in terms of wavelength. You hear terms like *a quarterwave vertical* or a *5/8 wave collinear*. In some designs, antenna sections can be stacked one above the other to increase the gain—to radiate more power. There are cases where only minimum gain is needed so that the radiated power does not interfere with neighboring signals.

So, what are we talking about in terms of dimensions? The length of the radiated wave of RF energy in the 800 MHz cellular band is about 12 in. long. To give some idea of how this dimension compares with other, more common frequencies and wavelengths, let's look at a typical FM and AM signal in the broadcast band. There is an FM station in the Washington, D.C. area that broadcasts on a frequency of 91 MHz. The RF wavelength at this frequency turns out to be a few inches over 10 feet long. An AM station broadcasting on 630 kHz radiates a wave that is almost 1500 ft long. You can see the comparison between the length of antennas needed for the AM and FM stations and the one for cellular radiotelephones.

In addition to the carefully calculated length, all antennas work better and with greater efficiency with a conductive ground at the base.

In the early days of radio, the term *ground* referred only to the earth or water over which the antenna was placed. For the best results, the ground should conduct the RF energy. The best natural ground is water and some broadcast antennas have been erected on piers driven into the ocean bottom or out in swamps. Where the composition of the soil was not particularly conductive, the ground characteristics have been improved by laying copper wire as a mesh or as radials under the base of an antenna.

In a cellular radiotelephone installation, the roof of the vehicle or the lid of the trunk serves as a ground. Installers prefer to mount the antenna on the trunk since it is easier to route the antenna cable or *lead-in* to the transceiver which is probably located right in the trunk. In a roof-mounted installation, the cable must be snaked down between the roof and the headliner and then back to the transceiver.

On cars that use fiberglass and other plastics for the fabrication of some of the body panels—the Chevrolet Corvette is a good example—a satisfactory ground for the antenna becomes a bit of a problem. One solution is to mount an aluminum disk or panel under the antenna inside the body or trunk lid. Another solution is to use a glass-mounted antenna on the rear window. To avoid drilling holes in the roof of the vehicle or the lid of the trunk, a glass-mounted antenna has been devised in which the antenna base is mounted high up on the windshield or the rear window next to the roof. The RF energy to and from the transceiver is transferred through the glass by means of a special coupling circuit contained in the antenna mount. There is no need to drill any holes anywhere in the vehicle.

Figures 6-15 through 6-18 show the antenna installation for the cellular radiotelephone that we have been following through this chapter. In Fig. 6-15, a 3/4 in. hole has been drilled through the trunk lid and in Fig. 6-16, the antenna has been inserted into the hole and the mounting nut is being tightened. The lead-in cable is being threaded down behind the trunk lining in Fig. 6-17. Figure 6-18 shows the completed antenna installation.

The installation in Fig. 6-19 shows an easily removed roof-mounted antenna being removed before going through a car wash. Note the configuration of the back of the car—there is no trunk lid in which to mount the antenna.

There appears to be little or no drop in operating efficiency of the antenna when comparing the trunk mount with the conventional on-roof mount. Radiation patterns made by an antenna used with these two

**Figure 6-15.** Drilling antenna mounting hole in car trunk.

types of mounting tested under carefully controlled conditions show almost similar results.

Earlier in this section it was mentioned that in some designs the antenna is made up of partial wavelength sections that are stacked one above the other. This is done to reinforce the energy. In discussions of antenna designs of this type, you will see references to statements like

**Figure 6-16.** Tightening antenna mounting nut.

**Figure 6-17.** Routing antenna lead-in to transceiver location.

**Figure 6-18.** Completed antenna installation.

3 dB gain which is a comparison between the power radiated from the antenna and the power radiated from some standard antenna design, generally a halfwave dipole.

Where there is no need for maximum gain from the antenna, a

**Figure 6-19.** Motorola's removable antenna. (Courtesy of Motorola, Inc.)

simple *unity gain* model is used. This means that there is no RF-reinforcing techniques used in the design. Situations in which this type of antenna would be appropriate are those in which the cells are small in size or where there is a possibility of interference with frequency channels in neighboring cells.

This discussion completes the description of a complete mobile cellular radiotelephone system. Now let's look at the portable and briefcase units and some of the problems encountered with cellular radiotelephones in general.

The cellular radiotelephone concept carried one step further means that the user need not be tied to the instrument in the vehicle. The technology that led to the development of small, battery-operated, hand-held communications units that are visible in use by security personnel can certainly be applied to the cellular radiotelephone. There are instruments available that can be used in the car as a conventional cellular radiotelephone and then unplugged from the cradle in the vehicle and used as a portable unit. Users are no longer limited by the length of the cord connecting them to the cradle—they can go anywhere.

A typical portable handheld unit, manufactured by Motorola, is shown in Fig. 6-20. This portable instrument has all of the features that have been described so far. The battery that operates the unit permits about eight hours of operation of which 20 to 30 minutes can be used in the transmit mode which uses considerably more power. Then it becomes necessary to recharge the batteries. A *rapid charge* brings the battery back up to full strength in an hour while the normal, recommended charging rate takes about ten hours. The usual method of operation is to have a spare battery on the charger overnight and to change to a fresh battery every morning. The battery for the portable unit can also be charged from a vehicular battery when the unit is not being used.

Where a longer battery life is needed such as at an emergency site in a remote area, in special events like parade, or at a construction site where a temporary phone is not feasible, a transportable unit is availa-

**Figure 6-20.** Using a portable cellular telephone. (Courtesy of Motorola, Inc.)

ble. This has a larger, greater capacity set of batteries that will operate the telephone up to 12 hours and, during that time, will permit about an hour and a half of talking. This unit is naturally bulkier and heavier than a handheld model and is about the size of a thick briefcase. These batteries can also be charged from any source or from a vehicle, and the cellular unit has a meter to show the state of charge of the batteries.

In addition to the handheld portable models and the somewhat heavier and larger transportable model just described, a number of companies are packaging instruments in business briefcases. Figure 6-21 illustrates the OKI model. In this unit, the briefcase lid is hinged so that the entire unit need not be exposed when the telephone is used—only the small flap needs to be turned back. The detachable antenna is carried on the lid at the hinges. The attaching straps can be seen in the illustration.

A more elaborate version is shown in Fig. 6-22. This model by Spectrum Cellular Corporation uses a standard aluminum case lined with leather and a smaller antenna mounted on the briefcase lid.

There is a growing use for these portable units for business persons or organizations in an area for a relatively short time. Typical of this type of use is a movie or TV production company that will be on location in the area and has a requirement for a lot of communication facilities for a few days or weeks. The need disappears when the work is completed and the group disbands.

This challenge and how to meet it is exemplified by an organization in New York that keeps 20 to 30 briefcase units in stock and in

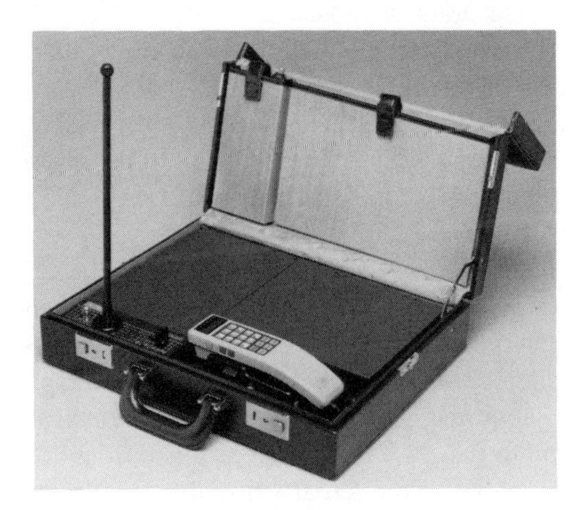

**Figure 6-21.** The OKI briefcase unit.

**Figure 6-22.** The Spectrum Cellular briefcase unit. (Courtesy of Spectrum Cellular Corp.)

use. The units are rented for $25 per day plus the air time used in calls. The business is expected to grow when a large convention center in the area is completed. It is believed that the exhibitors at the trade shows as well as visiting businesspersons have a need for short-term use of a telephone and these units would fit the bill.

Along this same line, recreational vehicle rental companies are equipping some RV models with cellular radiotelephones for people who need office or living space on a short term basis.

There is a price to be paid for the flexibility offered by portable units. The transportable unit described earlier, not an instrument mounted in a briefcase, will cost about 25 percent more than the conventional unit that goes in a car. The cost of the briefcase models will depend largely on the style and type of the case used.

The cellular radiotelephone, since its antenna emits radio energy, is one more source of RF radiation added to a spectrum already crowded with electromagnetic emissions. These electromagnetic emissions, in addition to the legal ones such as those from radio stations and installations, can radiate from a diathermy machine in the local hospital to a microwave oven in your kitchen, from a garage door opener to the cable that brings the programs to your TV set. As we depend more and more on microprocessors to measure, adjust, and turn on and off those things we need during our daily lives, there will be more and more chance for interference problems.

Your car probably has some kind of microprocessor that is programmed to monitor and adjust essential things such as the flow of the fuel in the fuel injection system, the advance or retarding of the spark, the cruise control, the anti-skid braking system. If you have a top-of-the-line model of one of the more luxurious cars, there is a microprocessor that is constantly monitoring, computing, and displaying the miles per gallon of gasoline that you are getting, telling you to buckle up your seat belt, and giving you a host of other information. All these functions are activated by RF energy generated by a microprocessor.

If you run into a problem with your cellular installation interfering with the operation of any of these electronically-controlled operations, you will not be the first. There was a long history of RFI—radio frequency interference—before the introduction of the cellular concept that probably began with the first mobile radio installation back in the 1920s. Radio amateurs (hams) are well aware of RFI since they get blamed for most types of interference. Some of the amateur handbooks have detailed chapters devoted to these problems and their solutions. There have been problems with CB radio installations too, but until the introduction of microprocessors into the control systems of automobiles, the problems were confined mostly to such things as the inadvertent activation of garage door openers or perhaps a little crosshatching or snow on nearby TV screens. Now the situation is different and, since the automobile manufacturers consider the installation of a mobile telephone as a modification to their car, the car warranty may not hold if you have an RFI problem.

The cellular equipment manufacturers are aware of the possible problems. They publish warnings in installation instructions and send out service bulletins. The major problems seem to occur when cables from the cellular unit are run too close to any of the car wiring carrying signals from the car's electronic control modules. Car wiring that is not shielded can act as an antenna and either pick up or radiate electronic signals whether it is supposed to or not.

In addition to care in the routing of the wiring, the transceiver/logic unit should be installed as far away as possible from any of the electronic control modules. In most installations this is easy since the unit is generally put in the trunk, away from the engine compartment and the wiring. The grounding wire used to connect the cellular radio to the car structure as a ground should be as short as possible and, again, should be kept as far away as possible from the car's electronic system. Another point to consider during installation is that the cellular antenna should be installed as far away as possible from the car FM/ AM radio antenna to lessen the chances of RF from the cellular radio getting into the car radio or tape installation.

The low power used in cellular systems is an advantage over other radiotelephone systems. A maximum of 3 to 5 w is used in a cellular radiotelephone while some of the other mobile radios will radiate from 30 to 50 w.

Electromagnetic interference has not been a major problem so far in cellular radiotelephone installations but you should be aware of the problems that can arise and discuss the possibilities with the technician that installs your cellular radiotelephone. Find out if there have been any problems with cars similar to yours and, if so, how were they cured. Check, too, with the dealer from whom you bought your car to see what the manufacturer has to offer in case of a problem. While you are doing this, ask if there is going to be any problems with the warranty on the car. The fact that a properly installed and maintained cellular system can be satisfactory is attested to by the continued growth of the cellular industry and increase in the number of cellular antennas seen on cars.

Having gotten a satisfactory installation in your car, the next question is how safe is it to use your telephone while you are on the road. It is fairly obvious that,in today's traffic, anything that takes the driver's attention away from driving and surrounding traffic conditions is a potential source of danger. The early cellular ads showing a happy, carefree businessperson driving along, holding the steering wheel with the left hand and a cellular telephone with the right, were not endorsed by those concerned with safety on the highways. The American Automobile Association (AAA) and the National Safety Council (NSC) have issued warnings about the use of a mobile telephone while the car is in motion.

The AAA cites a study by the Institute for Research and Public Safety that found driver inattention was a factor in nearly 25 percent of the accidents studied. Both the NSC and the AAA stress that if a call is to be made from the car it should be done when the car is not in motion. You should stop the car to dial the number. If your telephone rings while you are driving, you should carefully assess the traffic situation before answering the call to be sure that you can do so safely.

Equipment manufacturers as well as system operators recognize the need for safe operating and driving practices and make options available for the cellular installation that contributes to this. One option is a remote microphone/speaker installed on the sun visor so that the driver's eyes do not need to be taken off the road to carry on a conversation. Of course, you can use a remote speaker in the handset cradle so that the handset does not need to be taken from the cradle for operation. All these things help the driver operate the vehicle and still use the cellular telephone in a safe manner.

In September 1984, Ameritech Mobile Communications,the wireline cellular operator in the Chicago area and the first one to go on the air in the country conducted a survey of their customers. At the time the survey was made, Ameritech had more subscribers than any other system in the country—about 25 percent of the national total. This study showed that 4 percent of those surveyed had had an accident since their mobile radiotelephone was installed. This is half the national average of the drivers having accidents. In addition, the respondents to the survey said that they fastened their seat belts 45 percent of the time they were driving. The national average of those using seatbelts is 14 percent. This study seems to show that cellular radiotelephone users, as a group, are more safety conscious than the average driver.

During the extensive testing period before commercial introduction of the cellular telephone, Bell Labs did a study (1978) on driver behavior in conjunction with the use of a mobile phone using a controlled test group. Bell personnel reported that none of the subjects reported any difficulty in using a mobile radiotelephone. The use of the instrument in the car was compared with other driving tasks such as adjusting the heater or air conditioner controls, tuning the radio, looking for a landmark, and other typical chores. Answering the mobile radiotelephone and conversing in the car was found to be no more difficult than talking with another passenger.

Ameritech, in cooperation with NSC, published a folder entitled *Ten Tips For Safe Driving with Your Cellular Mobile Phone*. They made a direct mailing to all their customers and gave them to their retail outlets and sales offices for distribution to all their new subscribers. The hints given in this brochure are reprinted in Fig. 6-23. Figure 6-24 shows the tips sent out with the news release by the AAA.

Both Ameritech and AAA emphasize, however, the more positive side of having a cellular radiotelephone. They state how useful it can be

1. ***Purchase the proper car phone.*** *Select a cellular phone with the hands-free speakerphone option. Have the microphone installed on the sun visor directly above your line of vision.*

2. ***Assure correct installation.*** *Make sure the handset is installed for your driving comfort. The handset should be easily accessible, within comfortable reach, allowing you to sit and drive normally, and as close to your line of vision as possible.*

3. ***Be knowledgeable.*** *Insist on a demonstration of how to use your phone safely at the time the phone is installed or while on a test drive. Read the user's manual.*

4. ***Assess the traffic.*** *Before placing or receiving calls, assess the traffic situation and make sure you are fully aware of road and traffic distractions.*

5. ***Keep your hands on the wheel, eyes on the road.*** *Use your speakerphone, keeping both hands on the wheel, your eyes on the road and the phone handset in its cradle whenever the car is in motion.*

6. ***Call when stopped.*** *Dial phone numbers when your vehicle is stopped whenever possible. Remember your cellular phone is designed to allow you to enter telephone numbers and then send the call at your convenience.*

7. ***Use memory dialing.*** *Store frequently called phone numbers within the phone's memory to minimize dialing while driving.*

8. ***Driving is your priority.*** *Obey all traffic signs and signals and stay within the speed limits. Drive in the slow traffic lane in case you wish to pull over to complete your phone call.*

9. ***Minimize distractions.*** *Do not attempt to take notes while driving. Pull over first or use a pocket dictation unit. Prepare your phone calling list before you start out on the road.*

10. ***Buckle up.*** *Seat belts have been proven to save lives. Use your seat belt at all times and set an example for others.*

**Figure 6-23.** Ameritech's safety tips. (Courtesy of Ameritech Mobile Communications, Inc.)

for reporting traffic accidents and emergencies and other situations where the rapid dissemination of information can reduce aggravation and perhaps save lives. In addition, they show the need for the driver's concentration on the actual driving and the careful use of the cellular telephone.

Another question of safety is that of possible exposure of the user to RF radiation from the hand-held, portable units. With the increase of consumer and manufacturing usage of RF generating devices, the public has been made aware of the possible dangers of such devices and greater attention has been paid to the testing and labelling of such items. Consumer units such as microwave ovens come from the manufacturer with a safety certificate that discusses the precautions that have

AAA TIPS FOR SAFE USE OF CELLULAR CAR PHONES

1. Insist on a demonstration of how to use the phones safely during any test drive.
2. Select a model with hands-free microphone option which allows driver to talk without holding the handset.
3. Seek literature on how to operate this model safely.
4. Ask about security features of each model you consider. Your expensive investment will be attractive to others who can see it in your unattended car.
5. Install the microphone on the visor directly above your line of vision so you can drive and speak without having to turn your head into the microphone.
6. Keep both hands on the steering wheel and the phone handset in its cradle whenever the car is in motion.
7. Place calls only when the car is stopped. Use on-hook dialing and one-digit recall of frequently used numbers to speed the time it takes to place calls.
8. When receiving a call, assess the traffic situation before answering, then lift the handset briefly to stop the ring and replace the handset to continue the conversation on the hands-free microphone.
9. If talking on the phone while the car is moving cannot be avoided, drive in the slow lane. Keep the conversation brief.
10. If the conversation requires note-taking or complex thought, stop the car in a safe location or offer to return the call as soon as you can stop. The tractor-trailer in your rear-view mirror needs your attention first.
11. Whenever you use a cellular phone while driving, realize that you may be endangering yourself, your passengers, and other motorists.

Courtesy AAA Potomac

**Figure 6-24.** Safety hints from AAA. (Courtesy of American Automobile Association.)

been taken to prevent any possible exposure of the user to the RF energy that is used to do the cooking.

The hand-held, portable cellular radiotelephones, in the transmit mode, are the only instruments that have been questioned. Tests by Motorola, Inc., and the FCC working with the U.S. Food and Drug Administration's Center for Devices and Radiological Health, have shown that when the units are oriented as they would be when in use—with the antenna pointed away from the head and eyes—the minute dosage received by the user is within the limits set for safety by the American National Standards Institute.

In the receive mode the RF energy is so very low that there is no danger to the user. The RF power radiated in the other links of the telecommunications chain, from the cell site to the switch via microwave link for example, is inaccessible to the public and poses no threat.

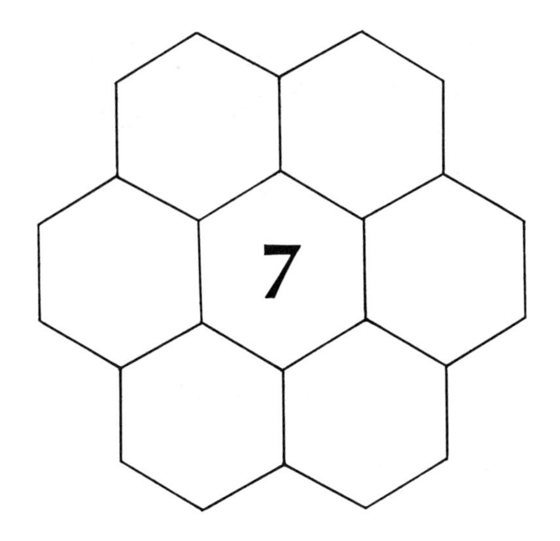


# The Role of the Regulatory Agencies

Earlier in the book, we discussed the beginning of communication without wires. Since the nineteenth century, the telecommunications industry has grown beyond Marconi's or anyone else's imagination. But in those early days, as the industry began to grow, conflicts arose between commercial and military users, the manufacturers, and the various groups of users. In those days there were no regulations on frequencies used or the amount of power pumped into the antennas that were sprouting up all over the country and throughout the world. Congress made several attempts to regulate radio communications in the United States, but it wasn't until the Radio Act of 1912 that things began to get sorted out in a serious fashion.

The Radio Act established some organization and order. Radio station and operator licenses were made mandatory and specific portions of the then known spectrum of frequencies were assigned to the various users. As more became known about this method of communication, and as advances were made in the equipment, the number and the needs of the users grew and further government intervention became necessary. New license requirements for both operators and stations were drawn up and new frequency assignments were made at the Radio Conference of 1924.

The growth of the industry continued at such a rate that in 1927 a new government agency, the Federal Radio Commission (FRC), was organized to oversee all radio communications. Until this time all the work had been handled by a section of the Department of Commerce, but now the FRC was an independent government agency.

However, this was not to last. Seven years later (1934) another communications act was passed and the role of the newly formed Federal Communications Commission (FCC) was expanded to oversee all types of wired and wireless communication. An increased role of the government and its regulatory agency was brought about by further increases in the communication needs of the nation for both business and personal use. New knowledge and technology made additional portions of the radio spectrum useable for increased frequency needs. Because of the growth of international communications in the post-World War I period, the FCC was asked to represent this country's interests abroad and to present and defend U.S. positions in international meetings concerning the uses of the radio spectrum.

It is fairly obvious why national regulations and international agreements are necessary. Since radio waves are no respecters of international boundaries, it became apparent that there was a need for international cooperation in the use of this very valuable but very limited resource. The International Telecommunications Union (ITU) headquartered in Geneva, Switzerland, is the body charged with coordinating the use of the radio spectrum among the countries of the world. This organization, like most international bodies of this type, has no actual police powers, but most nations recognize that it is in their own interests to agree to the uses of these frequencies and to abide by the international agreements. Periodic meetings of the member nations are held by the ITU to discuss any needed changes to existing agreements that have been brought about by technological improvements or by changing needs and conditions.

Unlike the longer wavelength radio frequencies used around the world for broadcasting and for commercial and government point-to-point communication, the energy radiated in a 800 MHz cellular system rarely goes beyond the line of sight. Because of this, there are few, if any, questions raised pertaining to cellular communication regarding international cooperation in North America. The U.S. has had a history of settling questions like this with its neighbors, Canada and Mexico, with little difficulty. In Europe, however, it is a different situation because of national chauvinism. (More about this in Chapter 9.)

Let's look now at the regulatory agencies, national and state, and at the part they play in the cellular business.

The FCC, as we have said, is the primary government agency that oversees the activities of the telecommunications industry. It depends upon your point of view as to whether the agency does a good job in refereeing between what appears to be two mutually exclusive concepts: the public interest and the profit motive. The drawing of the electromagnetic spectrum, Fig. 7-1, shows only the major divisions, a very small part that is the responsibility of the FCC.

To perform its assigned functions, the FCC, with its headquarters located in downtown Washington, D.C., has a supervisory and administrative staff and four operating bureaus: the Common Carrier Bureau, the Private Radio Bureau, the Mass Media Bureau and the Field Operations Bureau. In addition, there is an Office of Science and Technology that keeps the agency up to date on the state of the art and technical developments. See Fig. 7-2 for an organization chart. Fig. 7-3 shows the Commissioners in session.

The Common Carrier Bureau is concerned with common carrier communications methods, telephone companies, commercial radio other than broadcasting, satellite communications, and—the area that is of most interest to us—public land mobile communications.

The Private Radio Bureau oversees communications pertaining to aviation, public safety such as police and fire services, business, forestry, CB, and the hams or radio amateurs. In other words, they oversee any use of radio by a business whose activity is not providing any radio service commercially.

The Mass Media Bureau looks after the broadcasting end of the business. This includes AM and FM radio, TV stations, and the cable TV business.

The investigation and enforcement of FCC rules and regulations are in the hands of the Field Operations Bureau. They maintain monitoring stations and mobile monitoring units to listen and ensure that all types of stations abide by the terms of their licenses.

To do all of this work, the agency has about 3000 employees in the U.S. and at some overseas posts.

The FCC part in the cellular picture began formally in 1970 when they invited the telecommunications industry to submit proposals for new and hopefully innovative systems to improve and expand the existing mobile radio and radiotelephone service. As part of its overview of the mobile radio industry, the FCC had come to the conclusion that there was a definite need for a new approach and members of the industry agreed wholeheartedly.

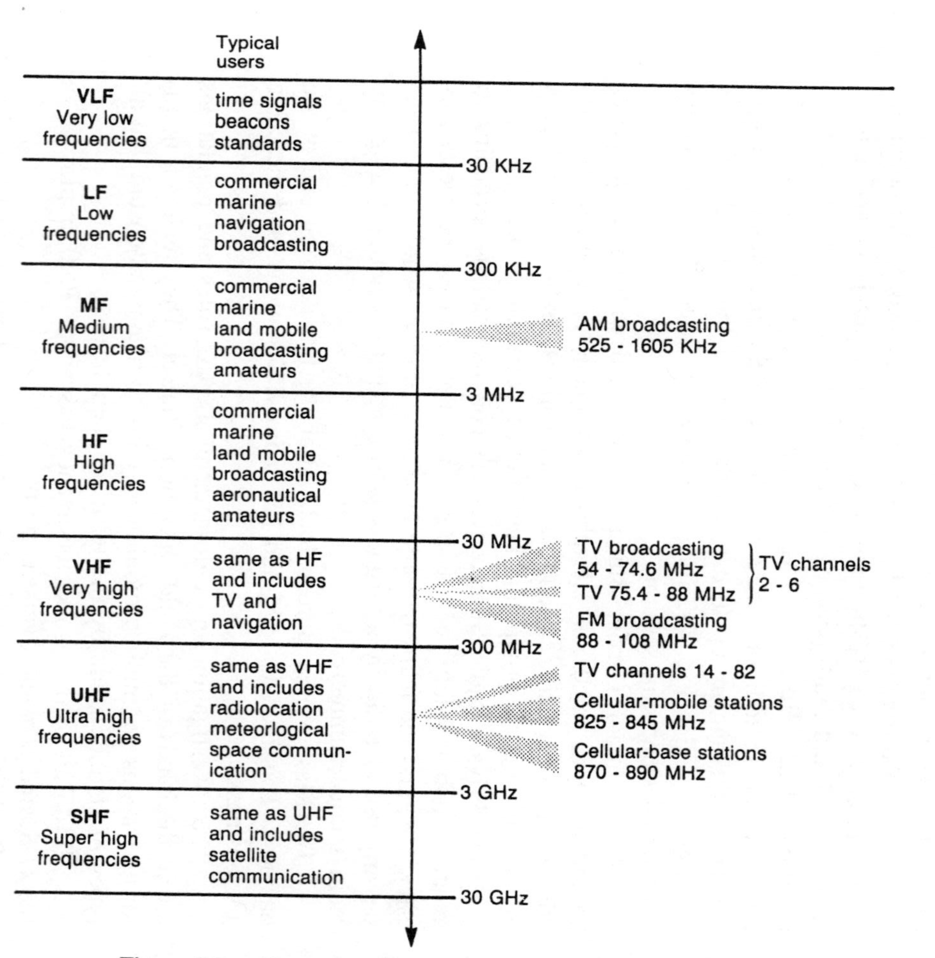


**Figure 7-1.** Electromagnetic spectrum used for communication purposes.

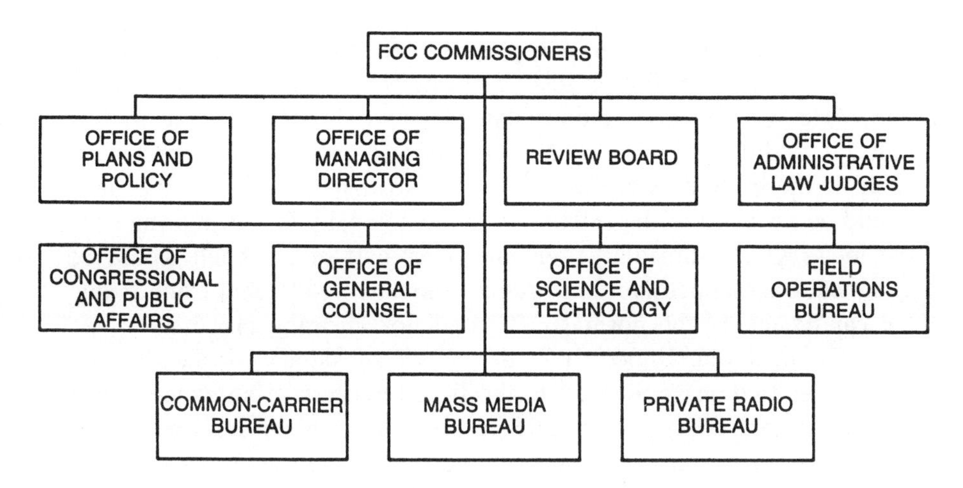


**Figure 7-2.** Organization of the FCC.

In 1971, in response to this request, AT&T filed a proposal describing a cellular mobile radio telephone system that could be located in the presently unoccupied 800 MHz region of the spectrum. The people at AT&Ts Bell Labs had been looking at this cellular concept for about twenty years and had tested and proved its feasibility.

**Figure 7-3.** FCC commisioners at a hearing. (With permission from *Broadcasting Magazine*.)

After studying the AT&T proposal, in 1974 the FCC allocated some frequencies for further testing of this new cellular system—825 to 845 MHz for the base stations and 870 to 890 MHz for the mobile units. In addition, if this system were to be adopted for commercial use each of these bands were to be divided so that one half of the band would be assigned to the existing wireline company in the area (namely, the local AT&T subsidiary) and the other half of the band would be assigned to a competitive nonwireline company be selected from among the candidates by the normal FCC regulatory hearing process.

The next year Illinois Bell, the AT&T subsidiary for the Chicago area, submitted an application to the FCC to construct an experimental cellular system covering the Chicago area and to begin tests on such a system. In 1977, the FCC approved the Illinois Bell application and in 1978, the system tests began with about two thousand customers as test subjects. During this same period a nonwireline company, the American Radio Telephone Service (ARTS) also had a cellular application approved and began testing a system in the Washington-Baltimore area. This system was constructed and operated under contract with Motorola who had designed and built the equipment they were using.

The FCC and the telecommunications industry was, of course, closely monitoring the results of these tests. In 1982 when it became clear that the testing was successful and that a valid market existed for cellular systems, the FCC finalized the tentative decision that two licensees would be awarded in each of the markets to be covered.

The term *market* or *markets* is one commonly used to define a specific area of the country. The definition used by the FCC follows pretty much the standard one used by Census Bureau statisticians, demographers, and urban planners, and is based on census data. When all the factors being considered reach a certain level, the area is designated as a Metropolitan Statistical Area (MSA), or New England County Metropolitan Area (NECMA). A Cellular Geographic Service Area (CGSA)—an FCC term—must be at least 75 percent of the area or population of the MSA or NECMA. This is the area that prospective licensees must define and provide system coverage for in the license application. The CGSA and the NECMA may or may not be the same as the MSA.

The markets were ranked into groups of thirty for consideration for cellular licenses beginning with New York as the first. In June of 1982, the FCC opened the 192 applications that had been filed for consideration for the markets one (New York) to thirty (Portland, OR).

At this point let's take a superficial look at the task the FCC staff was required to do with these applications to sort them out, review them, and then select what they considered to be the best applicant. This is the comparative hearing process and generally consists of four phases: (1) the predesignation, (2) the hearing, (3) the exceptions and (4) the final award.

The predesignation phase was broken down into three steps to be carried out by the staff of the Mobile Services Branch of the Common Carrier Bureau. Each application was officially received at the FCC headquarters and logged in. Each application was examined to determine if all of the required technical, financial and managerial information was supplied in sufficient detail to enable the examiners to properly evaluate the application. Last, the staff considered such things as a request for an exception to some particular regulation or a request for a waiver to some technical requirement because of engineering problems with that particular system or area of operation.

After the successful completion of these steps, many of which involved considerable correspondence with the applicant, the application was designated as a suitable candidate.

All during this predesignation phase, the applicant could file amendments to the original application to answer questions raised by the FCC staff during the review of the application, or to allow for additional technical information which had become available. Additionally, the bureau staff prepared an in-depth analysis of each application and made its recommendations.

The applications were then designated for a comparative hearing before one of the administrative law judges. The judge reviewed the record in detail, asked for further information if necessary, and eventually came to a decision to grant an award to one of the applications and deny awards to the others.

The hearing phase is supposedly the last phase in the process, the one in which the most qualified applicant is selected and the much-valued construction permit for the system is issued. However, this is not the way it usually goes because, in nearly all the cases that come before the FCC, one or more of the unsuccessful applicants will file an appeal or take exception to the FCC decision. They can ask for reconsideration by the commission or can go to the courts. The entire process can take over two years and be inordinately expensive, not only to the applicant, but also to the taxpayer since a large part of the FCC staff's time is taken up with this burdensome process. One applicant in the hearings for the first 30 markets estimated the cost to each applicant to be between $150,000 and $400,000 to do the necessary paper work and to hire expert witnesses to provide the additional engineering expertise and testimony pertinent to the application.

In an attempt to cut down on the time and expense involved in this licensing process, the FCC devised the lottery procedure for the issuance of succeeding licenses. This procedure also consists of three phases: a prelottery phase, the actual lottery and the postlottery phase.

The prelottery phase is about the same as in the comparative hearing process except that the so-called *petitions to deny* (legal points raised against any of the applicants), is eliminated. In the past this had been the most time-consuming part of the procedure.

When the prelottery phase has been completed and all the applications have finally been accepted or rejected, the FCC sets a lottery date approximately thirty days later. On this date all the successful applicants for that particular market are assigned a number which is chosen by a random process. The order in which the numbers are drawn establishes the ranking of the applicants.

After the drawing, petitions to deny can be filed against the winner by any of the unsuccessful applicants and the commission will consider these petitions. If the petitions are denied, the original winner is awarded the license and can go ahead with the preliminary construction of the system.

If the FCC review finds that there is merit in any of the points brought up by the petitions and the original winner's application is considered defective, the application is denied and returned, and the next ranking applicant is declared the winner. The review process is then repeated until the final grant is made.

In an attempt to further shorten the process by reducing the number of applicants to be considered, the FCC encourages *settlements* among those competing for the license prior to the drawing. The following is a relatively simple example—some of the later settlements got very complicated.

The American Radio Telephone Service (ARTS) began a test operation in 1980 of a cellular system in the Washington-Baltimore area with the equipment manufacturer, Motorola. In 1983, when the FCC opened the bidding for the permanent license for this area market, one of the top 30 in the country, ARTS made an application. Four other potential licensees also submitted applications. These were the Washington Post, a publishing and TV station conglomerate; Metromedia, a TV, radio and telecommunications organization; Metrocall, a paging company, and Metropolitan Radio Telephone System, a cellular system consulting firm.

The first step prior to the FCC decision as to the winning application was a merger of the interests of the Washington Post and ARTS. Then the resulting combination was sold to Metromedia for cash and a small percentage of the business. All the other applicants then joined the group and the resulting organization, operating under the name of Cellular One, got the license.

This settlement process has been so successful that the FCC had to actually hold a lottery in only three markets in the top 90—none in the top 30.

When very optimistic predictions began to appear in the trade press about the large amounts of money expected to be made in the cellular radiotelephone business, the number of applications that were filed for the MSAs ranked from 31 to 120 increased tremendously. About 1,000 applications were filed for the cellular licenses for markets ranked 31 to 90. For the much smaller, less populated areas (MSAs ranked 91 to 120), over 5200 applications were sent to the FCC.

There is another factor that contributes to the large number of applications that were sitting on the shelves at the FCC. While the original filing package in markets 1 to 30 might cost up to $400,000, the smaller markets that can use a simpler system—in some cases consisting of only one or two cells—a standardized system engineering package can be purchased for about $1,000 from any of a number of consulting firms. When this engineering package was put together with a vague statement as to where the money was going to come from for construction and operation, anyone could set up as a potential cellular mobile radiotelephone system operator.

The FCC, in accordance with the general *deregulation* policy of the Reagan administration, proposes to do very little regulating of the licensees after the license is granted. Any changes in the system within the CGSA, the area the licensee proposes to serve, will not require prior notice to the commission. This area is carefully defined in the engineering data that accompanies the license application, and the proposed location of the cell sites are chosen to be sure that the antenna radiation patterns cover the entire area with the useable RF signal strength required by the FCC. The operator can add to or change the location of the cell sites, for example, as long as the contours of this useable signal do not fall outside the limits of the CGSA. There is no hearing procedure to go through; the licensee merely has to notify the FCC that a change has been made.

If, however, the system operator wants to increase the coverage and extend the areas of the CGSA, an application is required and the request is subject to petitions to deny by interested parties.

For the individual state regulation of cellular operations, the role of the various state regulatory agencies is somewhat different. The states have traditionally exercised some degree of control over the public utilities operating within the state boundaries, usually setting rates and overseeing service. The degree of control depends, of course, upon the state legislature and the laws enacted.

Radio, however, because of the interstate and international characteristics of the radiated signal, comes under the control of the federal government. The FCC, as we have seen, awards the licenses and permits for construction and system start up for commercial radio, TV stations, and conventional and cellular mobile radiotelephone systems. The FCC has nothing to say, however, about the fees that the subscribers will be charged by the cellular company. This is the responsibility of the various state public utility commissions.

In 1984, the National Association of Regulatory Utility Commissioners conducted a survey of the state commissions to determine which states were planning to regulate cellular communications companies and to what degree. At the time of the survey, about 19 wireline companies in the top 30 markets were either on the air or close to it. Of these 19 areas, there were 7 competing nonwireline systems on the air or under construction.

Forty-four state utility commissions responded to the survey. The degree of regulation expected to be enacted by the state legislatures varies. Ten of the states plan to regulate both the FCC licensees and their resellers, while four plan to regulate only the licensee. Four others will regulate the licensee but as yet have not decided what to do about the resellers. There are 10 states that will not regulate the cellular companies in any manner at the present time and 16 have not decided what to do about this new communications phenomenon.

As time goes on and more experience is gained, there is sure to be a number of changes in the legislative picture. These changes will probably depend on the political climate at the time the subject comes up for consideration in the state legislatures. For the past five years deregulation has been the watchword in government circles and it has been fashionable "to let the market forces control" the regulatory process. As in many other things, only time will tell the final result.

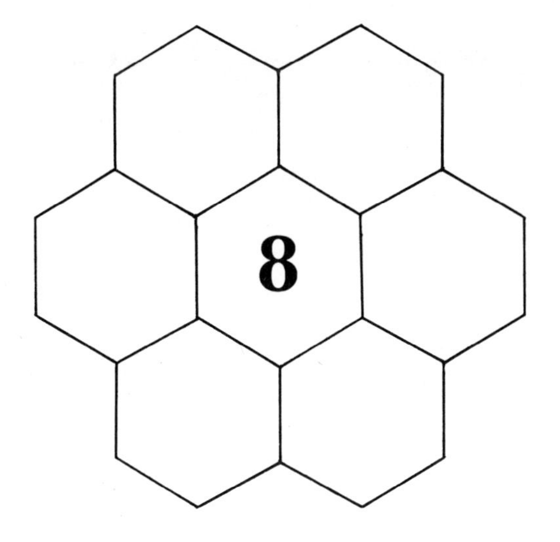


# The Economics of the System

As we said in the introduction, this is not a how-to-do-it manual on setting up and operating your own cellular mobile radiotelephone system. However, now that you have a reasonably good handle on all of the bits and pieces that make up the system and how they work, it might be a good time to sit back and look at the big picture.

One of the best ways to view this industry is to arrange all of the factors under consideration in tiers. In tier one are the actual providers of the service—companies like Bell Atlantic and Cellular One in the Washington-Baltimore area, NYEX in New York, Ameritech in Chicago, and PacTel on the west coast to give a small sampling. These organizations raise money to build the system, deal with the FCC and the local regulatory agencies, and build or contract to have built the actual cellular system. They also operate the system, round up dealers and customers, handle the billing of the users, and do most of the advertising.

The firms in tier two are the systems builders. These firms put the hardware in place, erect the antennas, and run tests to be sure that everything is satisfactory before turning the system over to the tier one organization. They have the expertise and technical know-how and can, if necessary, operate the system when completed. Some names in this field are Motorola, Ericsson, NEC, and AT&T.

Tier three consists of the manufacturers of the cellular instruments. The list of names is very long and includes Motorola, OKI, Toshiba, Mitsubishi, NEC, Panasonic, and AT&T. There may also be a number of private label instruments that are made by one of the bigger firms and marketed under another name much like Sears does with their household appliances and other equipment.

In tier four are two groups of retailers; resellers and agents. Resellers buy hours of operating time from the system operators (those in tier one) and resell this operating time to users in small increments, perhaps in blocks of 6 to 10 seconds. One of the users of this type of service might be a second competitive cellular company that received the necessary permit from the FCC but does not have their system running. The user of this competing system does not know or care whose equipment is handling the call. The monthly bill has no indication that the service was really provided by a competing system. In this type of arrangement, the resellers would get a discount for the bulk purchase of time.

The agents in tier four are selected by the system operators to install and maintain the customer's equipment, antenna and handset. They get a commission from the system operator and also make a profit on equipment that the customer buys. These customers have seen ads in newspapers or on TV or may have seen an instrument in a neighbor's car. They may have called the system operator for information and, hopefully, to sign up for that particular system. They will then be directed to an agent to have the equipment installed and the assigned cellular number burned in.

Tier five consists of people who do nothing but install and maintain the subscriber's instruments. Most of these firms are already in the mobile radio business and are familiar with the problems associated with this type of communication. These firms would install a cellular radiotelephone that was bought at a discount house or from an electronic supplier such as Radio Shack. They may also sell their own line of radiotelephones.

In tier six are the people who deal with esoteric matters such as the software needed to operate the computers and the switch that is the heart of the cellular system. Here, too, would be the management consultants that would help you set up the system efficiently and, if necessary, run it for you under contract after you are on the air. In this tier, would also be the engineering consultants who design the systems and the lawyers who handle the paperwork to secure the applications.

In tier seven are all the related industries and businesses that support the cellular industry. Here you will find firms which specialize in and build antennas for both the cell sites and the vehicles. Here you will also find the people who buy and package an instrument in a briefcase, and the people who build an interface that hooks up a personal lap computer to a cellular radiotelephone to transmit data.

The organizations that fall into tier one are the those that need high financing. Economists call this business a very capital intensive one. Each cell site can cost between $250,000 to $500,000 and the MTSO can cost that much as well. It is easy to see what the costs are likely to be before the first subscriber is signed up.

The profits that can be made from this investment are a little more difficult to discover. Figures for the wireline companies can probably be determined from the rate filings that have to be made to the state utility commissions and other regulatory agencies. Nonwireline money is hard to trace in most organizations because of the number of companies that have a piece of the action. Most of these companies are in other businesses too, and the money that comes from cellular may be combined with other funds in a stockholders' annual report.

But enough about the providers of this service. What is it going to cost you, the user, to have an instrument installed in your vehicle or boat and what will it cost you every month to keep it there?

Firm figures at this time may not be valid for very long because competition between the tier three companies has driven the cost of hardware down as more companies got into the market. There are other options available to the prospective buyer, varying in price, that require study and knowledge of the expected pattern of use before an economical decision can be reached.

One option that has been advertised well is the leasing plan. In the Washington-Baltimore area in the spring of 1985, you could find a plan that offered installation of the instrument in your car and 60 to 100 minutes of calling time for $99 a month. At the end of the lease period, generally about five years, you could buy the instrument for about 10 percent of the original cost. Leasing plans are available in most areas of the country and are probably used most by the business community.

In February of 1986, if you had a subscription fee of $250, you could rent an instrument for less than $20 a month. This included installation and insurance. At the end of three years, you could buy the instrument for a final payment of $250. In addition to these costs, there is a monthly charge for calls under whatever service plan you select.

Another alternative is to buy your instrument, have it installed and a number burned in from one of the cellular systems serving your area. The instrument and the installation can cost $1500 to $3000 depending on how many bells and whistles are attached.

There are a number of service plans available and the proper choice will depend upon what you want the telephone for. Here are a couple of typical examples of what you can expect when you begin to look at these plans.

What is termed a *regular use plan* will cost $35 a month subscription fee. In so-called prime time (7:00 AM to 7:00 PM), the first minute or fraction of minute of use will cost forty cents and each additional tenth of a minute will cost four cents. During nonprime time and on Saturdays, Sundays, and holidays the rates are twenty-four cents and approximately two cents respectively.

There is also a *package plan* an *occasional use plan*, a high use plan, and an off-peak plan. The off-peak plan costs $10 a month for the subscription plan which includes 100 minutes of nonprime time use. Time is billed in one minute increments with the additional nonprime time rate of fifteen-cents a minute and prime time rate of sixty cents a minute (or fraction of a minute).

When considering these plans, don't forget that, unlike your home telephone, you will also be paying these same rates for the incoming calls.

You can see that you will have to think about how you are going to use this convenience and then study the various plans offered in your area to see which one fits in with your expected pattern of use. If you can, talk to people who have cellular radiotelephones in their cars. Find out about the quality of service of the system they are connected with, particularly if there are competing systems in your area. If you are going to be driving outside the area covered by the local system(s), find out about roaming agreements with the neighboring systems.

Look at a map of the system coverage to find out if you live on the edge of the coverage. If you do, have the salesman drive you out there to see how the system works at the outer limits. Is it consistent? Is the audio quality OK? As you can see, buying a cellular radiotelephone is not the same as calling up the local telephone company and having an instrument installed in your house. To get a mobile cellular radiotelephone installed so that it operates efficiently will take time and a lot of questioning and reading, but it will be worth the trouble and effort.

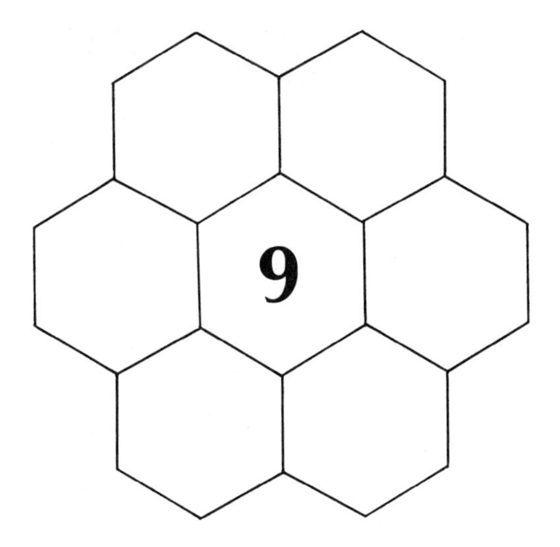


# What is Being Done Elsewhere

As we have seen, the cellular concept offers a solution to some of the telecommunications problems of the densely populated, highly industrialized countries of the western world. Here is another example. On many of the inhabited islands along the coasts and fiords of the Scandinavian countries, travel by boat is essential and is as common as travel by automobile in the U.S. Expense and technical difficulties have hampered the expansion of the telephone lines to these islands beyond the mainland. The installation of a cellular system for use by portable and fixed stations at home and by mobile units in boats and cars expands the communications horizon for these citizens.

For lesser developed countries the cellular concept offers a method of linking together those population centers that are too far apart to economically connect by wire. No country can develop its resources and grow successfully without a reliable means of communication available to all its citizens. In the U.S., this came about through the invention and popularization of the telegraph and telephone.

To be completely satisfactory, the telephone system must be simple, inexpensive, and flexible enough so that it can be used in both urban communities as well as the less densely populated rural areas. A cellular system appears to meet most if not all of these requirements. Single cells covering up to 20 miles across in sparsely populated areas or small urban centers can be linked together by microwave relays.

This eliminates the need for laying cables, erecting poles, and stringing wire. In some types of terrain, the conventional construction methods are impossible; swamps or river deltas are obvious examples. Many towns in tropical countries are built on such ground.

There is an added flexibility to cellular in that the radio frequencies used for communication can be selected according to the conditions that will be encountered. Low frequencies, 400 MHz for example, will cover longer distances and give better foliage penetration than the high frequencies of 800 or 900 MHz. This is important in tropical conditions where jungles and rain forests are frequently encountered. This advantage could not be used in the U.S. because by the time the FCC got around to looking at cellular and authorizing its use, the 400 MHz band was already filled with other users. Too, foliage penetration is not a major consideration here.

A cellular system is certainly quick and easy to set up and put in operation. Nearly everything can be preassembled prior to installation. Furthermore, in contrast to conventional wireline systems, a battery bank kept charged by solar cells is feasible because of the low power involved. The years to come—not too many of them—will show whether full advantage is being taken of the features that a cellular system offers.

Figure 9-1 shows a compilation of what is known about systems operating in the world today. Plans for expansion, of course, will depend on the economic picture in that particular country or area. Oil producers will have trouble financing new construction during times when the price of oil is low. The same problems exist in lesser developed countries when interest rates go up and loans have to be refinanced. In cases like this a nationwide cellular system is a luxury and generally has to wait for better times.

In Europe, however, economic and technical development have paralleled the U.S. In telecommunications, land mobile radiotelephones have been in use for years using the same call-handling techniques as are used in the U.S. The systems began to suffer from the same problems as those in the U.S. described earlier—lack of channel capacity and nowhere to go in the radio spectrum. The existing systems could not accommodate future and waiting customers.

In small, nationalistic countries such as those in Western Europe, technical problems are not the only problems that the telecommunications industry faces. The political differences that have plagued the European Common Market since its inception carry over into the cellu-

| Country/ Service Area | Frequency (MHz) | Maximum Subscriber Capacity | Number of Sub-scribers | Date Service Started | Number of Cells | Maximum Channel Capacity | Equipment | | | Operated by |
|---|---|---|---|---|---|---|---|---|---|---|
| | | | | | | | Switch | RF Base | Mobiles | |
| **AUSTRALIA**<br>Melbourne<br>Sydney | 500 | <br>4,000<br>4,000 | <br>3,000<br>3,000 | <br>Sep. 81<br>Dec. 81 | <br>4<br>7 | <br>120<br>120 | All NEC equipment | | | Australian Telecommunications Commission |
| **BAHRAIN** | 400 | 1,500 | 250 | May 78 | 2 | 20 | All Matsushita/Panasonic equipment | | | Bahrain Tel. Co. |
| **CANADA**<br>Alberta<br>(inc. Edmonton & Calgary) | 420 | 10,000 | 1,000 | Feb. 83 | 86<br>(no handoff) | 40 | Nearest Class 5 tel. office | G.E., NovAtel & others | NovAtel | Alberta Government Telephones |
| **HONG KONG** | 800 | 13,000 | 2,200 | June 83 | 14 | 800 | All NEC equipment | | | |
| **JAPAN**<br>Tokyo<br>Osaka<br>Nagoya<br>Fukuoka<br>Sapporo<br>Hiroshima<br>Sendai | 800 | <br>100,000<br>100,000<br>100,000<br>20,000<br>20,000<br>20,000<br>20,000 | <br>18,812<br>8,317<br>2,207<br>1,007<br>477<br>406<br>582 | <br>Dec. 79<br>Nov. 80<br>Jan. 82<br>Dec. 82<br>Dec. 82<br>Mar. 83<br>Mar. 83 | <br>88<br>58<br>38<br>22<br>14<br>23<br>14 | <br>1,000<br>1,000<br>1,000<br>1,000<br>1,000<br>1,000<br>1,000 | NEC | NEC | NEC<br>Matsushita/ Panasonic | Nippon Telegraph and Telephone PC |
| **MEXICO**<br>Mexico City | 400 | 8,000 | 2,600 | Aug. 81 | 1 | 80 | All NEC equipment | | | Telefonos de Mexico |
| **QATAR** | 400 | 5,000 | 3,250 | Feb. 82 | 13 | 400 | All Matsushita/Panasonic equipment | | | Ministry of of Comm./Trans. |
| **SAUDIA ARABIA**<br>Riyadh<br>Jeddah<br>Damman | 450 | 200,000 | 4,500 | Sep. 81 | 20 | Depends on System Growth | Ericsson<br>(3 switches) | Ericsson, Philips | Ericsson, Philips | Kingdom of Saudi Arabia PTT |
| **SCANDINAVIA**<br>Sweden<br>Norway<br>Denmark<br>Finland | 450<br>(gradual conv. to 800 MHz to begin in 1986) | <br>100,000<br>150,000<br>50,000<br>50,000 | <br>41,262<br>34,435<br>26,525<br>15,183 | <br>Oct. 81<br>Nov. 81<br>Jan. 82<br>Mar. 82 | <br>192<br>260<br>51<br>97 | About 10,000 | Ericsson<br>(6 switches) | Ericsson<br>Mitsubishi,<br>Mobira | AP (Phillips)<br>Ericsson<br>NEC<br>Mitsubishi<br>Mobira (Finnish)<br>Motorola<br>Panasonic<br>Siemens (German)<br>Simonsen (Norwegian)<br>Storno (GE)<br>Cancall (Danish) | The 4 PTTs in Sweden<br>Norway<br>Denmark<br>Finland |
| Totals: | | 350,000 | 117,405 | | 600 | | | | | |
| **SINGAPORE** | 400 | 6,000 | 6,000 | Nov. 82 | 4 | 180 | All NEC equipment | | | Telecommunications Authority of Singapore |
| **SPAIN**<br>Madrid | 450 | Will be extended to entire country | 350 | mid-82 | 2 | Depends on System Growth | All Ericsson Equipment | | | Spanish PTT |
| **UNITED ARAB EMIRATES** | 400 | 10,000 | 4,000 | May 82 | 33 | 400 | All Matsushita/Panasonic equipment | | | The Emirates Telecommunications |

**Figure 9-1.** Commercial cellular systems in operation in world today. (Courtesy Personal Communications Technology)

lar field. The problem is not helped by the fact that all the communication facilities are under government control except for England's recent action. In the years to come, it appears that there will be about four systems in use in the various countries, none compatible, in spite of the efforts of the Conference of European Postal and Telecommunications Administrations (CEPT), a multinational coordinating committee. Incompatibility is very inconvenient to businesspeople and holiday makers who use cellular radiotelephones, are accustomed to crossing international borders with little or no effort, and want to use their mobile telephones as they move. As far as cellular is concerned, the most that has been done so far on the continent is to agree to use frequencies of 809 to 915 MHz and 935 to 960 MHz with 25 kHz spacing for the ultimate cellular mobile radiotelephone system when and if the countries agree upon a suitable design.

Meanwhile, let's look at what is being done at the present time by some European countries to meet the increasing demand for mobile radiotelephones.

The first system developed in this area was the Nordic Mobile Telephone system (NMT), covering the Scandinavian countries of Denmark, Norway, Finland and Sweden. The specifications for this system, which operates in the 450 MHz band, were drawn up by a committee from the member countries' telecommunications departments.

NMT got on the air in 1981 and now has over 200,000 subscribers. Not only is the system used for conventional mobile customers but, in an area where a big portion of the population use boats to travel to and from work and for recreation, the cellular radio has been extremely adaptable and useful.

While the NMT system has been running successfully, it does have some drawbacks. Since it has been in existence for some time, it cannot take advantage of new developments and so it is less sophisticated than the new systems operating in the 800 and 900 MHz bands. It does not use the 450 MHz portion of the spectrum efficiently. Also, the NMT system does not have the capacity of the newer systems, nor can it be enlarged easily. Because of these drawbacks, the designers of the original system devised a new NMT which will work in the 900 MHz band.

In France, the External Telecommunications Department of French Telecom is beefing up the existing conventional land mobile radiotelephone service to meet the increased demand. Until the CEPT standard comes into use, expected in the 1990s, France and Germany have an agreement for the joint deployment of a cellular system called S.900. This system will use frequencies in the 900 MHz band.

The French portion of the system will go into service beginning with Paris, then Lyon, then through Alsace to Strasbourg, down to Marseille and Nice on the Italian border and then north to Calais and Lille. The map in Fig. 9-2 shows the intended plan. The highway net between these cities will also be covered by cells so that communication will be constant from Marseille in the south to Calais in the north

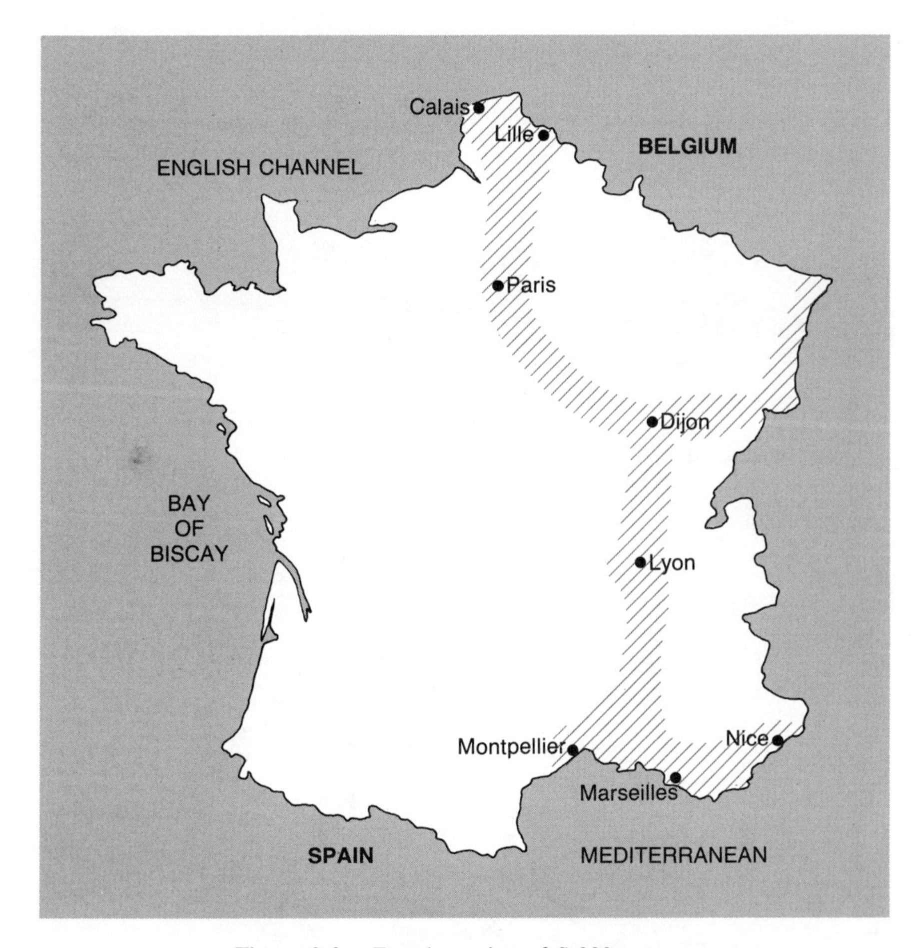


**Figure 9-2.** French portion of S.900 system.

on the English Channel and then through Strasbourg to the German system.

French Telecom expects to be serving 50,000 subscribers by 1991 by means of the S.900 cellular system and the more conventional radiotelephone system. At this time it is expected and hoped that a CEPT system will be ready for use throughout Europe.

In 1984, the Federal Republic of Germany through the Deutsche Bundespost introduced a national cellular service referred to as Netz C-450, on the 450 MHz band. The system coordinated with French Telecom is called Netz D-900. The map in Fig. 9-3 shows the coverage expected with this system. As with the French plans, this is a linking of the major urban areas.

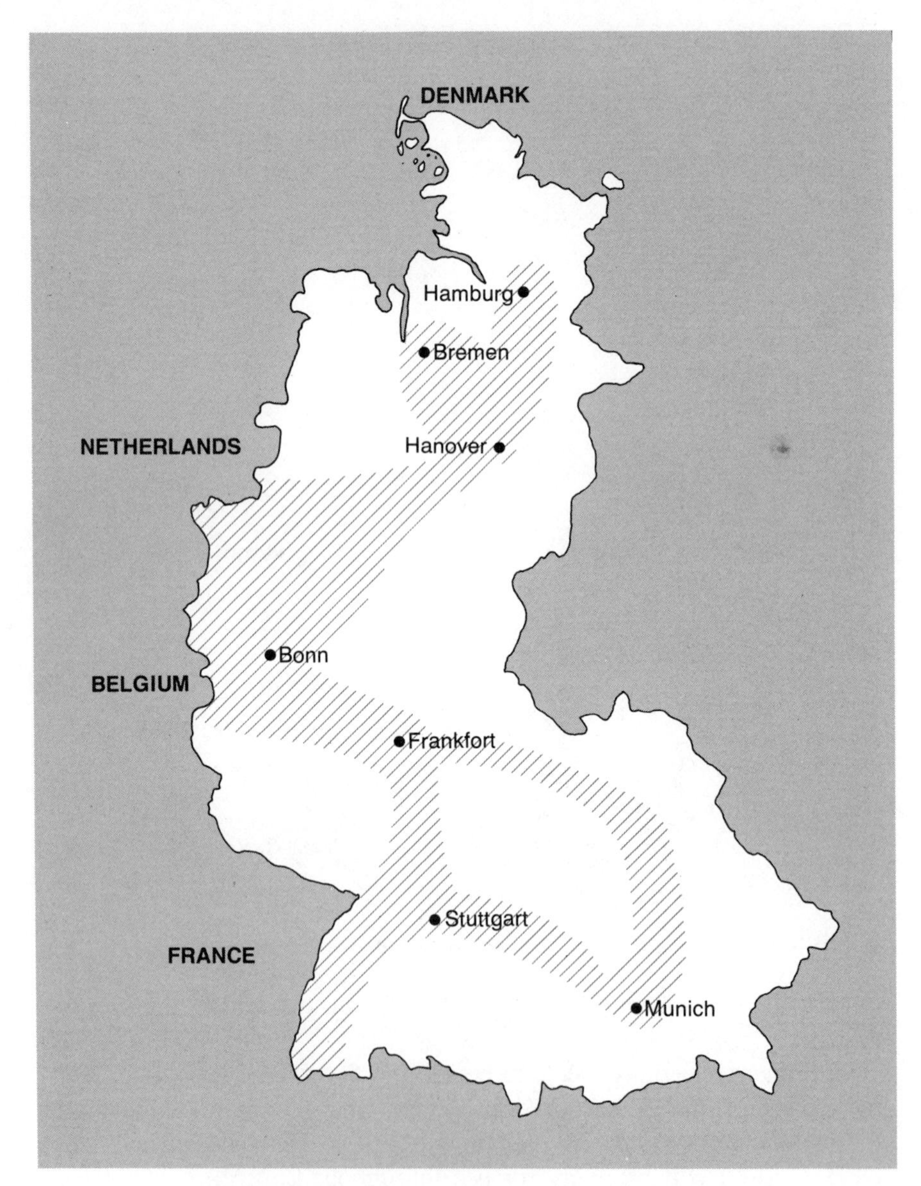


**Figure 9-3.** Coverage of West German cellular system.

Other European countries including Austria, Belgium, Ireland, the Netherlands and Spain are planning to adopt NMT pending the introduction of the CEPT concept. Italy has not yet made a decision as to what to do.

In the U.K. the situation is somewhat different. It is necessary to look at recent history to get the entire picture. In 1981, the British Parliament did some reorganizing and passed the Telecommunications Act. This act eliminated the old General Post Office (GPO), and separated the responsibilities of the postal functions and the telephone and telegraph functions. The wireline side of the old GPO was renamed British Telecom (BT). In 1984, BT passed into public ownership under the *privatization* schemes of the conservative government. Ownership was split up with 51 percent of the stock sold to the public and the remainder retained by the government.

During this period, decisions were made regarding the introduction of cellular mobile radiotelephones. The system selected was an improved AMPS, the name of the original system tested in Chicago, and renamed Total Access Communication System (TACS). It operated in the 900 MHz band.

Like the U.S. two competing systems cover the country including Northern Ireland. One system called Cellnet is 51 percent owned by BT and 49 percent owned by a private organization called Securicor. Securicor is a conglomerate with interests in security, cash transfer and coach (bus) building. More importantly, the company was active in the land mobile radio field prior to the organization of Cellnet.

The competing system has been named Vodaphone and is 80 percent owned by Racal, a telecommunications company; 15 percent owned by Millicom, a paging and telecommunications company with interests in the U.S. as well as the U.K.; and 5 percent owned by Hambros Bank.

According to the terms of the licenses issued by the government, both systems were to be operational by January 1, 1985 and available to 60% of the population by the end of that year. Both licensees met these requirements. By the end of 1989 both systems are to cover the land area in which 90% of the population is located. Figure 9-4 shows the planned coverage for the Cellnet system. The plans for Vodaphone are similar.

There are some different aspects between the cellular markets in England and the U.S. Because of the density and location of the population, and the greater reliance on public transportation system and walking particularly in city centers, there was expected to be more emphasis on portable and briefcase units than on mobile units. Experience has shown this to be true. Estimates are that there are about 60,000 mobile and portable units in operation at the present time.

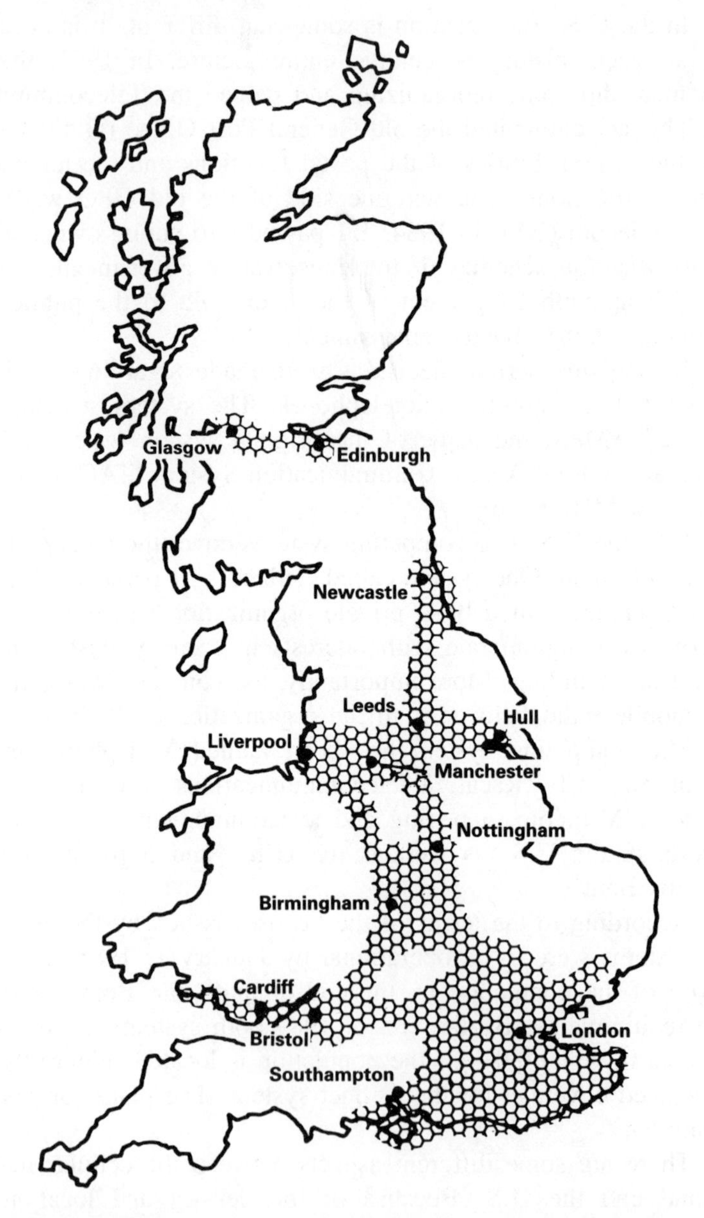


**Figure 9-4.** Proposed coverage of the Cellnet system. (Courtesy of Cellnet.)

The other point is that, because of the tax situation in England, company-furnished cars are considered part of nearly all midlevel technical and managerial jobs. About 60 percent of all new car sales in England are made to companies for this purpose. A company car is expected to be furnished in all positions in the salary range of $15,000 to $20,000 a year. It is obvious that a cellular radiotelephone will be an additional incentive so this market is expected to be the prime target for the first year or so.

There are other features planned to be available to the cellular-equipped British motorist. Using the facilities of the existing cellular systems, these will include up-to-the-minute traffic information using data furnished by the police and public works departments. After sorting and organizing the information in a data base program, the information will be routed to the appropriate cell site where it is pertinent and be available to the motorist on call on a liquid crystal display (LCD) display or a voice synthesizer. Also under investigation is the use of the same cellular network as a gross vehicle location system tracking the movement of a truck or car from cell to cell much as is envisioned in the U.S.

Japan has had a conventional cellular system in operation for about seven years that now has over fifty-thousand subscribers. Of more interest, though, is the personal radio service (PRS), which might be called the "poor man's cellular system."

The CB phenomenon was not known in Japan as it was in the U.S. The Japanese CB sets were limited to 8 channels and 1/2 w of power as compared to 40 channels and 5 w of power permitted in the U.S. by the FCC.

The PRS system, operating in the 900 MHz band, has 79 channels for the users and one control channel. With five watts of power, the typical range is four miles in urban areas and six miles or more in more open areas. The units are licensed in the same manner as are the CB sets in the U.S. A form packed with the set by the manufacturer is filled out by the user and mailed to a government agency that then issues the license.

The user then selects a personal ID number and enters it into the set when it is installed in the vehicle. This number, just like a conventional telephone number, must be known to anyone wanting to make a call to the user. When a call is made, the ID number of the person wanted is punched into the set via a conventional keypad. The process to complete the call then goes on automatically. The transceiver scans the 79 talk channels until it finds a clear channel. Then a *calling and control transmission* is sent blindly—there is no certainty, of course, that the wanted party is available. This transmission consists of the ID number of the calling station, the channel number used, the ID number of the person called and the control data necessary to complete the call.

If the wanted party is within range of the transmission, that person's unit, as it scans the talking channels, will recognize the ID and move to the vacant channel given in the control and calling transmission. The conversation can then begin.

There are some very definite advantages of the PRS over the current CB system in use in the U.S. There are almost twice as many channels (79 compared to 40), and the system is selective. This means you can make calls to a specific person and you do not have to listen to idle chatter while the set is on.

The PRS system has a so-called universal mode in which the user can get general calls from anyone who wants to chat without having a specific person in mind. This mode is similar to Channel 19 on CB and "CQ" on the ham radio. This channel can also be used for advertising messasges.

Of the 750,000 PRS sets licensed by the end of 1984, most were mobile units but portable hand-held units were becoming increasingly popular.

Cellular systems in other parts of the world, like most technological advances, are expected to grow to meet the needs of the economy. When the inhabitants can afford to install a cellular system, there will be any number of organizations that will be glad to design, install, and if necessary, operate it. History has shown that an efficient communication service is essential to a nation's growth, and cellular systems are part of this service.

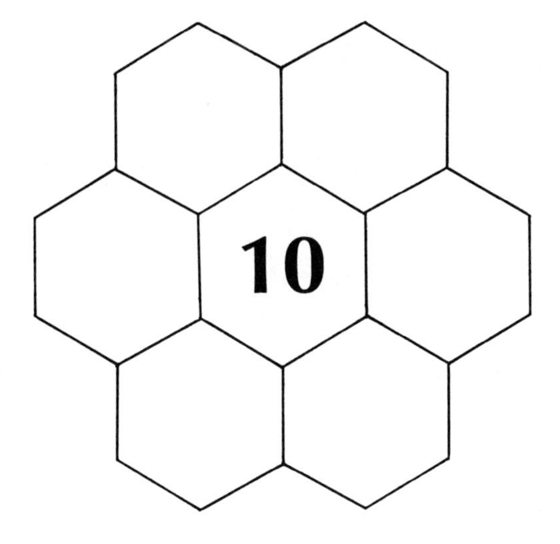


# The Future

There are concepts being considered, ideas being tested, and hardware being constructed to expand the uses of a cellular system for the consumer. Some things will not be of much use to the subscriber who uses the system only for the most obvious reasons—keeping in touch with the office or home. Other things may tip the scale in the decision between buying a cellular phone and doing without one.

This chapter contains a brief description of a few things under development. Some of them may never get implemented because they won't work satisfactorily or the market for them will not develop. Other things may be extremely useful to every cellular user and may become standard equipment in every installation. Still other ideas not yet thought about may become a great value to the present cellular user or bring new users into the market.

## Security

One of the things that can happen with a cellular telephone, without the user knowing anything about it is that the conversation can be overheard. It is more difficult to overhear a conversation on a cellular system than on a CB or conventional land mobile radio since cellular frequencies change as the user drives from one cell area to another and the transmission is handed off. Also, the frequencies in use can be any one of the hundreds of frequencies that have been assigned to the service. However, a professional in the industrial espionage field or a hobbyist who just likes to listen to whatever is on the air knows the frequencies in use in any particular area are on the FCC master list. This list is available to the general public either through magazines and newsletters that cater to this hobby or through one of the companies in Washington that research FCC documents and sell the information legally. Also, a scanner purchased at a local electronics store and programmed to scan the frequencies in use in that area will quickly pinpoint any calls. (See Fig. 10-1.)

The simplest, least expensive solution to the problem of privacy is to simply not discuss sensitive or private matters on your cellular radiotelephone. It might be well to inform the other party in the conversation that you are talking from your car before topics come up that you don't want overheard. There is generally no way anyone can tell from the quality of voice on the telephone that the call is coming from a car. Just consider that a call from your car will be about as private as talking in a restaurant or any other public place.

If it is important to you to be able to talk freely about private matters on your cellular phone, there are techniques and equipment available to make your conversations virtually listener-proof. The more security you want, the more it is going to cost. Also, whatever is done to your cellular instrument must also be done to the cellular instrument

**Figure 10-1.** A typical scanner that can cover the cellular frequencies. (Courtesy of Regency Electronics, Inc.)

of the person with whom you are carrying on the conversation. This problem is relatively simple to solve for public service organizations, police and fire departments, and government agencies that need privacy. The number of instruments they use is not very large so they can foolproof every instrument used in the system. Since you cannot do this, you must remember to whom you are talking and whether they have the same degree of security that you have.

Besides avoiding sensitive topics, there are some methods of insuring a reasonable degree of privacy for your calls. Scrambling the conversation so that it is garbled and unintelligible to any eavesdropper is probably the most practical method. There are several ways this can be done as illustrated in Fig. 10-2. The simplest and therefore the least expensive way to scramble a conversation is with a frequency-inversion scrambler which costs about five hundred dollars a unit. This package fits over your handset—it is about the size of a carton of cigarettes—and is coupled acoustically to the mouthpiece. The circuitry in the unit inverts the frequencies of your voice—the high tones become low and the low tones become high. Your voice sounds like a distorted Donald Duck to anyone listening without a scrambler. However, because there are not that many encoding frequencies, anyone with a scrambling unit, some determination, and time can soon find the code that you are using.

Another method of scrambling is the band splitting technique in which the voice is divided into bands of frequencies which are then changed and transmitted.

For about ten times the cost of a frequency-inversion unit you can get more security with a time-division scrambler that breaks the sound into very small sections, shifts them around into a different order, and then reassembles them at the listener's end to make the sound intelligible. There is a slight delay while the sound manipulation takes place but it is hardly noticeable to the listener.

The next step up the scale is a digital encryption unit which costs about twice as much as a time-division scrambler. However, it is probably the most secure method. In this technique, the audio information (your voice) is converted to digital form, that is, into ones and zeros. These bits of data are then rearranged according to a program that is in the encryption unit. The process is reversed, of course, at the other end of the circuit. You can see that there are a vast number of ways that the digits can be moved around and so it would be very difficult for a industrial spy (for example) to unscramble your conversation. At the

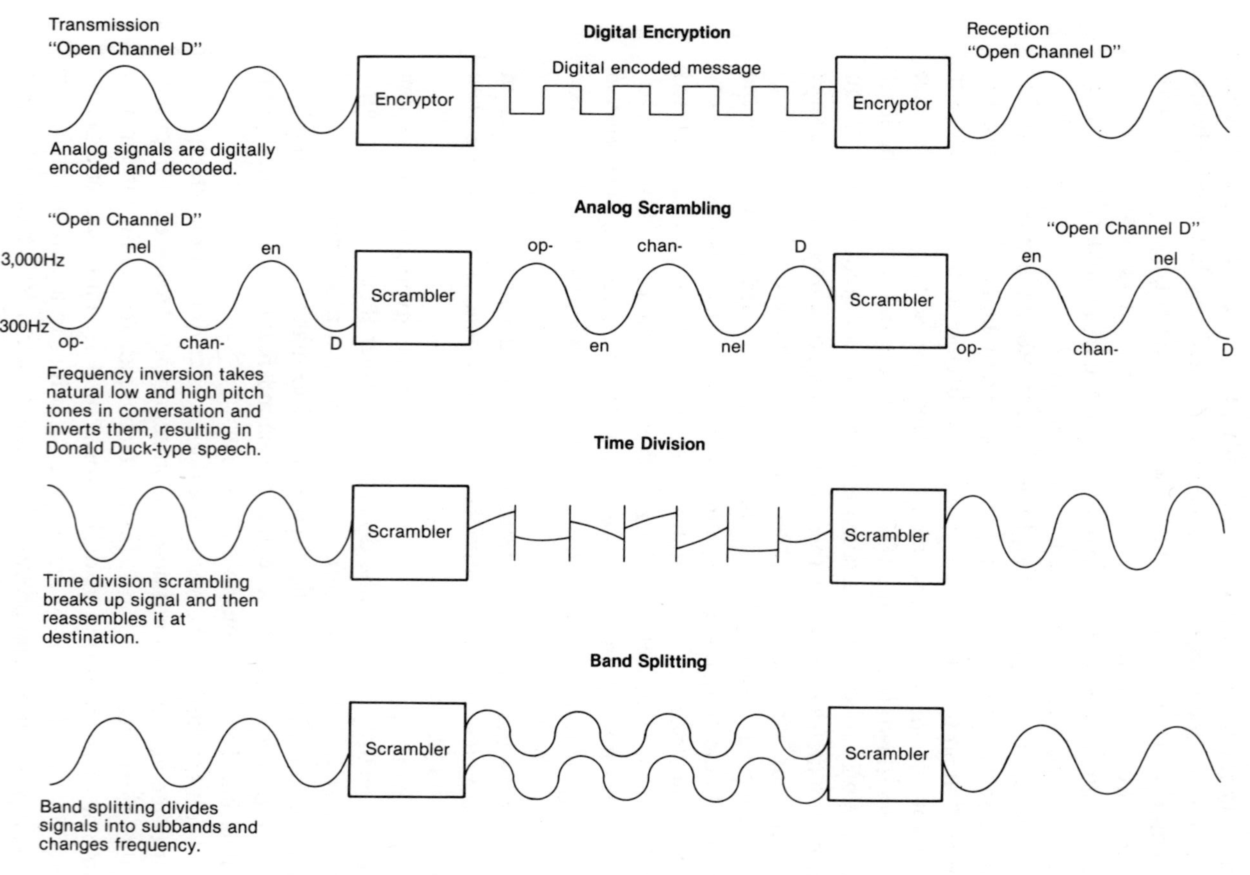


**Figure 10-2.** Scrambling methods. (Courtesy of *Cellular Business Magazine*.)

present time, however, the voice quality using this technique is degraded because high quality digital transmissions need a broader bandwidth—more space in the spectrum—than is now available in the current cellular systems.

## Digital

In Chapter 4 we briefly mentioned the difference between analog and digital signals. The illustration in Fig. 10-2 shows this difference. The circuitry of a digital system takes the analog signal (your voice) and chops it into bits which can be transmitted at a faster rate than the usual analog signal. The circuitry at the receiving end of the conversation reassembles these bits into the original analog form which we can recognize.

What is the advantage of digital signal? A digital signal takes up less space in the spectrum than an analog signal does, except when extremely high fidelity is wanted. The circuitry needed to handle a digital signal can be physically smaller, cost less, and not need as much power to operate. Things can be done with the digital signal that are impossible with analog signals. Any extraneous noise or interference that is picked up during the transmission can easily be filtered out and the signal bits reconstituted at the receiving end in exactly the same form as they were. The privacy problem, too, is greatly lessened. The digital technique is already used for transmission of data in other fields, from one computer to another for example. This use could be adapted for use in cellular systems. To date it has been done experimentally.

With all these advantages, why isn't this technique in use at the present time? The main reason, of course, is that the existing system (analog FM) is in place and working reasonably well. Also, the voice quality of a digital signal is not equal to the quality which telephone and cellular customers are used to, except when more space in the spectrum is taken and more elaborate equipment is used. This is not a factor in military communications systems where digital is used extensively but it is an important one in the civilian market where the customer is paying to hear recognizable voices.

The use of digital techniques will probably come to the cellular radiotelephone and if not, some other technique will evolve because of the need to utilize every kilocycle of the spectrum to accommodate the growing number of users.

## Data Handling

The transmission of data from a computer, large or small, is familiar activity for the wireline telephone company. Couplers of various types are available to enable a journalist, for example, to type his story on a portable computer in a motel room at or near the scene he has been sent to cover. The data from the computer goes via a coupler between the computer and the telephone in his motel room directly to the editorial room at his newspaper. Sales people using the same type of equipment can access a data bank to retrieve some needed information to help complete a sale.

A cellular system offers a telephone without wires but, in handling the data stream from a computer, the 100 millisecond break in communications that occurs when the mobile station's signal is transferred automatically from one cell site to another (the handoff) causes a problem. The computer that is churning out the data at a very rapid rate does not pause during this break and so some data bits do not get transferred.

One piece of equipment that has been designed to overcome this problem is *The Bridge* built by Spectrum Cellular Communications (see Fig. 10-5). The bridge, which is just a little smaller than a cellular handset, is connected between the computer and the vehicle's trans-

**Figure 10-3.** Transmitting computer data via "The Bridge." (Courtesy of Spectrum Cellular Corp.)

ceiver. When the transceiver gets the signal that tells it that a handoff to another cell site is going to be made, the bridge begins to store the data coming from the computer rather than send it on to the transceiver. When communication is restored through the new cell site, the stored information is transmitted.

A companion unit is needed for the fixed station at the other end of the circuit. Spectrum Cellular Communications calls this unit *The Span*, and it performs essentially the same function as the bridge.

## Cellular in a Fixed Service

It is obvious that there is no reason why a cellular radiotelephone system originally designed for mobile service could not also be used as a fixed telephone system. After all, if an automobile with a cellular installation is parked somewhere off the road and a call is made, it is performing as a fixed system. There are some applications developing that are natural extensions for cellular radiotelephones.

One application is the use of a cellular instrument as an emergency call box. Putting this instrument on a pole beside the highway, for example, means that there are no wires to string up or maintain. The New York City police department is purchasing about 750 cellular units to use throughout the city. Since these units will have only one application and need to make only one connection to the police department, there is no need for the conventional 12-button keypad on a handset. There is no need for the handset, either, that is subject to vandalism. The microphone and speaker can be mounted behind a tamper-proof grill.

The system can be programmed so that the location of the unit that has been activated—this takes just the press of a button—is given on a cathode ray tube (CRT) monitor at the police headquarters or answering station. There is no need for the caller to look for identifying information on the call box or for a stranger to try and locate themselves by looking for street signs or similar cues. This type of installation lends itself for use in places where emergency communication is most likely—public beaches and parks, stadiums, and so on come to mind at once.

Extending this concept still further, there are systems existing and under development that include a mobile MTSO. If needed because of the area to be covered, there are also portable cell sites. Systems such as this can be moved into an area to provide instant emergency communication in the event the local telephone system is out of action because of civil strife or natural causes such as hurricanes, earthquakes, and so on. The MTSO can be linked with outside communication lines by satellite or microwave.

A cellular radiotelephone system can be installed, too, in rural area or mountainous terrain where customers are located far apart and the cost of stringing wire for a conventional telephone system would be extremely high. The same holds true for operations such as oil rigs in remote areas or in water out of touch with land facilities.

## Position Locating Systems

The increased interest in vehicular communications for other than purely commercial reasons brought about by the introduction of the cellular mobile radiotelephone has sparked ideas for the development of other uses for the mobile radiotelephone other than just conversation. There is no reason why the mobile radiotelephone cannot be used for the transmission and reception of other information in the same manner as the conventional landbased wireline telephone systems. One of these uses has already been discussed—the transmission of data to and from a personal computer located in a vehicle. Another use is for a position locating system, essentially landbased navigation systems that will tell the vehicle's occupants just where they are and show them on a displayed map. An alternate configuration can be used as a monitor at some fixed location to keep a security staff aware of where a vehicle is located at all times.

To date, these systems are based on the use of an existing radio navigation system (Loran-C) developed some years ago for use by ships at sea and along the coast. Using low frequency radio signals from high powered landbased systems operated by the U.S. Coast Guard, the position of the ship or vehicle is determined by the difference in the arrival time of the radio signals from three widely separated stations. The received signals containing the pertinent information are processed in a small microcomputer that is a component of the Loran receiver. For ship navigation use, the location is displayed in latitude and longitude but it could be displayed in any other form.

Since the Loran-C radio signals already cover approximately 90 percent of the land area of the U.S., it was natural to turn to an inplace working system for adaptation to mobile use in a landbased method. One such commercial vehicle-locating system, the Auto-Trac, consists of a Loran-C receiver mounted in a vehicle and connected to the vehicle's cellular radio transceiver through the specially designed unit, the bridge. The Loran receiver processes the signals from the Loran transmitters and produces the location. This information is sent through The Bridge and the cellular system to a central computer. The Bridge, you will recall, allows the transmission of data without any missed bits of data due to the break in transmission during handoff.

The central computer upon receipt of this position information searches the files and displays the map of the appropriate area and the position of the vehicle on the monitor. The monitor can be at a headquarters used for a vehicle monitoring system. The information can also be transmitted back to the vehicle via the cellular telephone and displayed on a monitor screen mounted in the vehicle. Motorola developed a similar vehicle location system using conventional two-way radios to transmit and receive information.

The uses for such a system are many and varied. One of the first that comes to mind, particularly in today's political climate overseas, is the protection of diplomats or business executives while driving or being driven. A vehicle so equipped could be under the constant monitoring of security forces when a kidnapping or bombing threat is a possibility. A foot-operated switch could activate a warning signal at the security headquarters without the knowledge of anyone outside the vehicle. The same idea of constant or intermittent monitoring can be applied to the trucking industry. A deviation from the prescribed route, obvious to those at a headquarters monitoring station, would indicate that a highjacking operation or something other than normal business had taken place. The truck's exact location could be given to authorities for action.

While the market for installations such as this may be small, it is an extremely valuable weapon for public safety services. It is a convenience for others.

## The Mobile Satellite Service

With the growing body of technical knowledge, it has become possible to use a satellite as an antenna platform in a mobile radiotelephone system. The Mobile Satellite Service (MSS) is a system in which one two or more satellites serve as the support for the system antenna. Although there is somewhat more to it than just this. In a conventional two-way radio system, the satellites replace the tower used to support the land-based antenna. In a cellular system, the satellites replace the cell site that has the equipment to transfer the signal to the MTSO and the telephone company. A simplified drawing of a typical system is shown in Fig. 10-6.

The equipment on the ground at the receiving station would not be much more complicated than the existing cellular or land mobile equipment. Depending upon the radio frequencies used, the antennas for mobile use would be somewhat larger and more complicated but they could still be easily mounted on a vehicle. Proposed designs are illustrated in Fig. 10-7.

At last count there were about twelve proposals before the FCC for systems from organizations wanting to get into this business which can be expensive. The published estimates of the cost of an operating system range from $40 million to $700 million.

There are three general markets for this type of service. There is the commercial market that includes people whose needs are now being at least partially met by cellular or land mobile systems. Examples are police and fire departments and dispatching services. A second market includes the telephone users in rural or remote areas where this service would be less expensive than placing poles and stringing wire. It is estimated that 25 percent of the population of the U.S. lives outside a

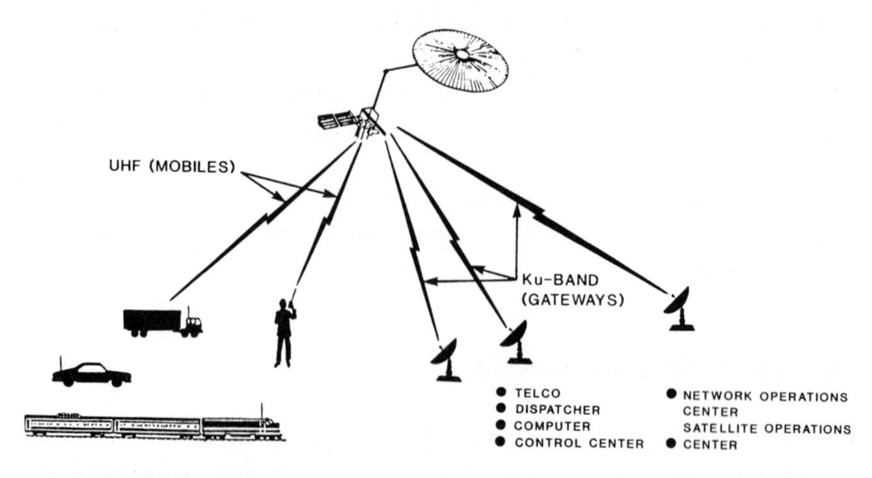


**Figure 10-4.** A concept for a mobile satellite service. (Courtesy of Mobile Satellite Corp.)

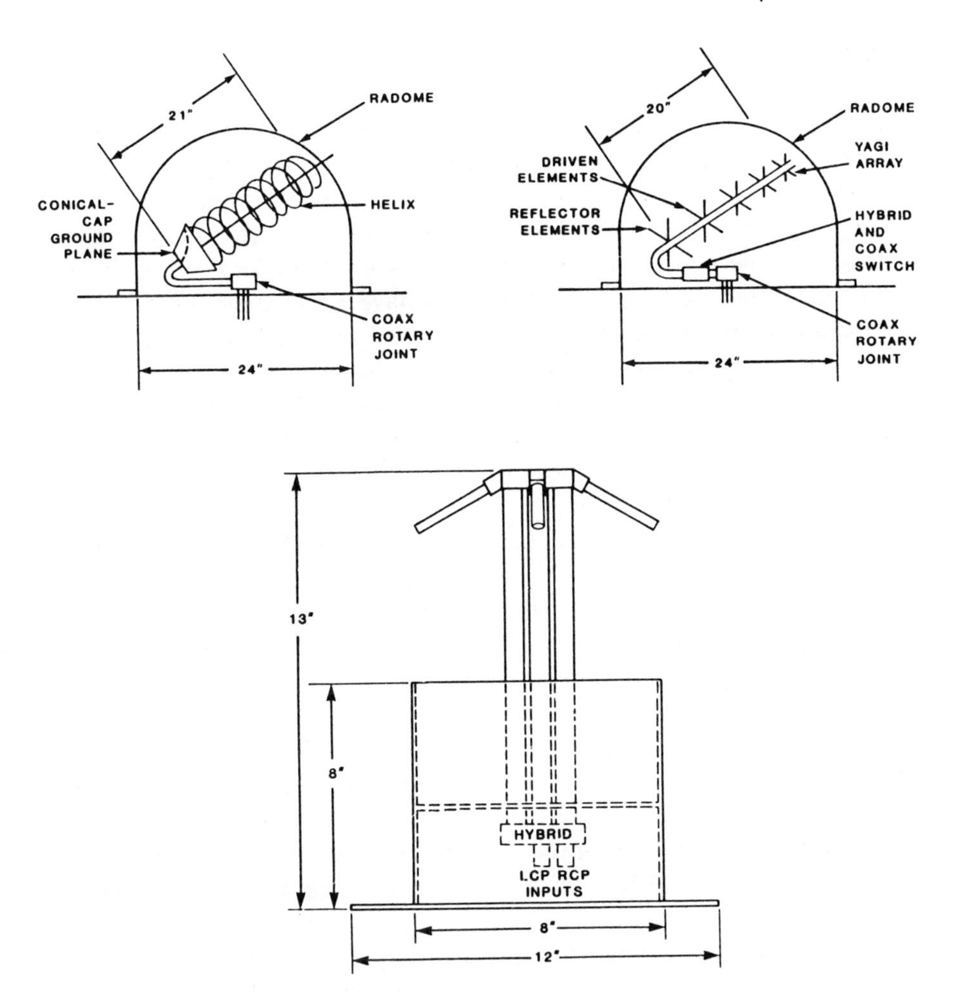


**Figure 10-5.** Types of antennas suitable for mobile mounting. (Courtesy of Mobile Satellite Corp.)

standard MSA and 20 percent of the industrial activity in the U.S. takes place outside of a standard MSA. The third market includes the new industrial users whose needs cannot be met by land mobile service or by cellular systems. Examples of this type of user would be long-haul trucking lines and off-shore oil rigs.

To date, the FCC has not decided whether the new service should be a monopoly operated by one or two companies or a wide-open service that anyone can get into if they have the money and an open slot on a satellite.

## Cellular PBX

The requirements for communication facilities in a modern office building have become much more than just needing a place to plug in a few telephones. There is a need to hook up an office Private Board Exchange (PBX) system (probably computerized). There are probably connections needed for antenna dishes for satellite systems, main frame computers, and personal computers that have become as necessary on the modern executive's desk as the calendar and the telephone. Architects and designers talk about *intelligent buildings* in which the telecommunications facilities are built in, along with the plumbing and electricity, while construction is under way.

All these facilities so necessary to the modern office system, are still dependent essentially upon a piece of wire to connect it to the system. True, there are the cordless telephones that have become popular for use at home, but the range of these instruments is limited restricting their use to the immediate area. This restriciton is no longer true for portable and mobile cellular radiotelephones; therefore why not use the existing cellular telephone system in the office environment. The major reason is that conventional cellular systems do not work well in buildings—the exterior and interior building walls act as a shield and keep the RF energy from coming in to the instrument or going out to the cell site.

However, each floor of the building or office area could be treated as a small scale cell. The equivalent of the cell site would be connected to the local telephone system. Each individual would be able to move about the building at will and never be out of range of the telephone. Moving an occupant from one office to another or just moving a desk would not involve moving a telephone connection or waiting for a new telephone number to be assigned. The original number would be good anywhere within the building or office system. Such a concept is shown in Fig. 10-8.

A system such as this is not without its problems, though. Among these are the possibility of lack of privacy, interference with other systems and the fact that RF could leak from the building in sufficient quantity to cause interference with other telecommunications systems inside or outside the building.

Why cannot be existing cellular system covering the area in which the building is located be used? The shielding effect is the main reason and this depends upon the composition and thickness of the walls and

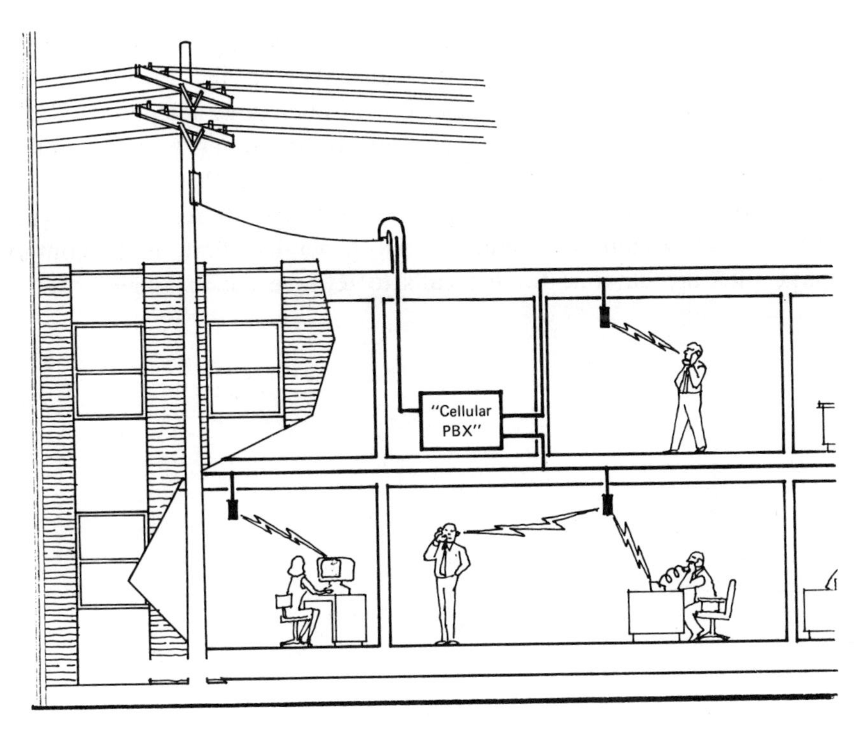


**Figure 10-6.** What a cellular PBX system might look like. (Courtesy of Personal Communications Technology.)

interior partitions. Since the effect of these factors is unknown and difficult to determine before tests can be carried out, it is best to assume that the signal from the existing system is not sufficiently dependable and cannot be used.

Proponents of the intelligent building, however, do not see these problems as unsolvable. Where security and privacy are necessary, the use of scrambling or other techniques could be used.

## Voice Mail

In an attempt to offer additional services for their subscribers, some of the cellular companies are offering what is known as *store-and-forward* or voice mail. This amounts to a verbal replacement of the pink slips that pile up on desks when the occupant is away—"Mr. Jones called and said . . . ," "Please call Mr. Smith any time after 4,"—you get the idea.

These voice mail systems are essentially storage places for messages that can be retrieved by the subscriber at his convenience by using the cellular instrument or any other Touch-Tone telephone. These systems can be elaborate and can forward messages to any number provided by the subscriber or notify him or her that a message is waiting by using a beeper. Charges for this type of service will vary from one organization to another but generally there is a monthly charge plus the air time for the calls to retrieve the messages.

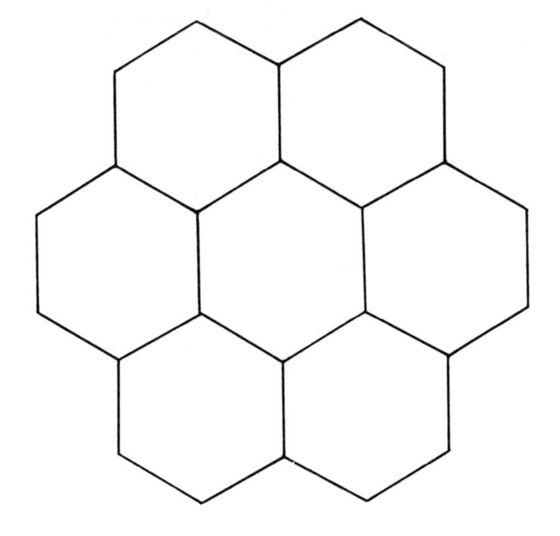

# Glossary

| | |
|---|---|
| AMPS | Advanced Mobile Phone Service. The name for the original cellular system tested in Chicago in 1984. |
| ARTS | American Radio Telephone Service. The original name given to the Washington/Baltimore system. |
| AXE | The symbol for the Ericsson switch or MTSO. |
| Bridge | A commercially marketed electronic unit that enables a cellular radiotelephone to handle digital data. |
| BT | British Telecom. The U.K. wireline company. |
| Burned in | A common term for putting additional data onto an already existing integrated circuit. Generally done in a service facility rather than at the factory at the time of manufacture. |
| CB | Citizen's Band. A group of frequencies at 27 MHz for private use with a maximum of 5 watts of power. System has a limited range. |
| Cell site | The link between the subscriber and the MTSO containing the transmitters, receivers and antennas for communication with mobile and portable stations and with the MTSO. |
| CCITT | International Telegraph and Telephone Consulting Committee. A worldwide organization set up to solve common telecommunications problems. |

| | |
|---|---|
| CEPT | Conference of European Postal and Telecommunications Administrations. Essentially the same as CCITT but confined to the European countries. |
| CGSA | Cellular Geographic Service Area. The area covered, or designed to be covered, by a cellular system. |
| Channels | A pair of radio frequencies assigned and used for communication so that simultaneous talking and listening can take place. |
| CRT | Cathode Ray Tube. A vacuum tube with a face similar to a TV screen on which electrical and RF data can be displayed. |
| CTIA | Cellular Telecommunications Industry Association. The industry trade association. |
| dB | Decibel. A term used as a measure of power. |
| DC/DC | Direct Current to Direct Current. An electronic circuit for raising or lowering the voltage of an incoming electrical direct current. |
| DTMF | Dual Tone Multi-Frequency. The generic term for the dialing tones used in the Touch-Tone dialing system. |
| EMXD | Electronic Mobile Switch. The Motorola MTSO |
| FCC | Federal Communications Commission. The federal regulatory agency that oversees all telecommunications activity. |
| FM | Frequency Modulation. A system commonly used for high fidelity radio transmission with additional characteristics that are useful for a cellular system. |
| FRC | Federal Radio Commission. The predecessor of the Federal Communications Commission. |
| GPO | General Post Office. The British postal service that formerly had responsibility for the telephone and telegraph service. |
| Hams | Amateur radio operators. The origin of the term is lost, but it is still widely used. |
| Hand-off | The automatic switching of a subscriber's call from one cell to another to insure serviceable communication. |
| IC | Integrated Circuit. A small solid state device on which a number of miniaturized electronic components are combined. |

| | |
|---|---|
| ID | Identification. A commonly used abbreviation. |
| IMTS | Improved Mobile Telephone System. An improved coventional land mobile system in which the subscriber can dial the wanted number directly. |
| ITU | International Telecommunications Union. The international coordinating body. |
| Landline | The local telephone system. |
| LCD | Liquid Crystal Display. An electronic method of displaying data on a flat surface. |
| LSI | Large Scale Integration. A method of packing a large number of components into an integrated circuit or chip. |
| MSA | Metropolitan Statistical Area. A geographic area having a number of common interests. Used in statistical and economic studies. |
| MTSO | Mobile Telephone Switching Office. The interface between the cellular system and the local telephone company. This office contains the switch. |
| NCS | Network Control System. The name for the ITT/E.F. Johnson switch. |
| NEAX | The Nippon Electric Company's name for their switch. |
| NECMA | New England County Metropolitan Area. A term used to denote an MSA in New England. |
| NMT | Nordic Mobile Telephone. The cellular system used in Scandinavia. |
| Off-hook | A term taken from the telephone industry which means to initiate a call. Literally, "to take the receiver off the hook." |
| PROM | Programmable Read Only Memory. The integrated chip that is put into the subscriber's transceiver at installation containing the individual identification and calling numbers. |
| PRS | Public Radio Service. An improved CB-type of radio communication as yet not approved by the FCC. |
| PTSN | Public Telephone Switched Network. The local telephone company. |
| PVC | Polyvinylchloride. A synthetic material used in the fabrication of piping as well as other items. |

| | |
|---|---|
| RCC | Radio Common Carrier. A company licensed to sell radio communication facilities and services much like the telephone company sells telephone services. |
| RF | Radio Frequency. A common term for the energy generated in radio circuits. |
| RFI | Radio Frequency Interference. Interference from one radio telecommunications system to another or to other electronic devices such as television. |
| Roaming | The act of moving from one cellular system covering a CGSA to another, separate system. |
| RSG | Remote Switch Group. An intermediate switch sometimes used between a group of cells and the MTSO. |
| rx | Receiver. An electronic shorthand term. |
| SMSA | Standard Metropolitan Statistical Area. An urban area defined by such factors as population, income, manufacturing, etc. |
| SPC | Stored Program Control. A telephone industry technique for using a computer to control the operation of the telephone system. |
| Switch | The computer that controls the operation of the cells and is the interface between the cellular system and the telephone company. |
| TACS | Total Access Cellular System. The improved AMPS system installed in the U.K. |
| Telco | Another name for the telephone company. |
| tx | Transmitter. An electronic shorthand term. |
| Wireline | A local telephone system. See also landline. |

# Addendum

Since work was begun on the preparation of this book the cellular mobile radiotelephone industry has undergone some natural changes. Some of these changes are associated with the organization and business side, and some with the technical side with the equipment and system design. The basic structure of the industry doesn't seem to have changed. Although the earlier excitement that took place in the lotteries and in the awarding of the licenses has simmered down, there is still a lot of enthusiasm for the industry as indicated by prices that have been paid for on-the-air systems. This addendum is an attempt to bring the information up to date and indicate which way the industry will probably go in the years to come. In some three years the industry has become a billion-dollar part of the national economy, and this figure is expected to increase tenfold by 1993 according to one industry observer.

As of the middle of 1986 service is available in over 90 markets across the country, with competing services on the air in about half of these areas, and permits issued and construction under way in the remainder. Nationwide there are now some 500,000 subscribers, with the growth now at a somewhat slower rate, which is to be expected.

That the industry has matured is indicated by the activity in the resale market of licenses and existing systems. The system or license price seems to be on a "per customer" or "potential customer" basis. One system in one of the larger markets sold for about $25 per potential customer, and there were rumors in the trade press of another sale that went for twice that figure. In another transaction, $1.65 billion was reported to be paid for a group of eight licenses. This market, nationwide, is somewhat complicated by legal ramifications. The FCC has a hand in the transfer of the licenses as do the courts, since these transfers generally come under the jurisdiction of the courts as part of the AT&T breakup of several years ago. And, of course, the state regulatory agencies have some control.

The use of the cellular telephone by the travelling subscriber has been made easier as more and more roaming agreements have been reached between neighboring systems. And for the industry, the CTIA is moving forward on the establishment of a nationwide clearing house to simplify the book- and record-keeping for those systems whose users travel and use their cellular telephones. Agreements such as this must step carefully to avoid any antitrust problems.

On the technical side, the efficiency of the equipment and the systems has undergone a natural improvement. The manufacturers and operating people have learned more about the systems, equipment, and customers and their pluses and minuses. This is a normal result of time on the air. More has been learned about the expansion of the systems, both in the land area covered and in the handling of additional customers within the same number of cells.

For the user, in addition to the less obvious system improvements, there have been improvements in the design of the instruments. Most of them are now lighter and less bulky and there has been an emphasis on the use of "hands off" instruments in vehicular-mounted installations for both increased safety and convenience. There are portable units now coming on the market that are half the size and weight of the original units, and is expected that the increase in roaming agreements will swell the number of users of this type of instrument. More subscribers will be walking around with their instrument clipped to their belt or carried in a handbag or briefcase.

Instruments have been developed specifically for maritime use since most of the US harbors and coastal waters are now covered by cellular systems. These units are essentially the same as the vehicular models in operation but have been redesigned and repackaged to cope with the adverse conditions likely to be found on the water.

The future of the cellular industry has already been discussed and nothing has appeared on the horizon since the writing of the earlier words to change any of the projections or prophecies.

# Index